S.S. Publishing

Naming Triangles

Name each triangle by its angles and sides.

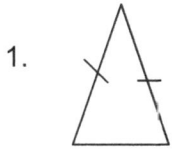

1.   isosceles , acute

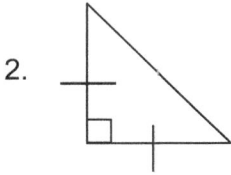

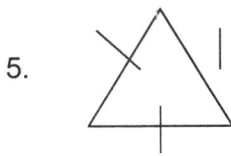

2.   isosceles , right

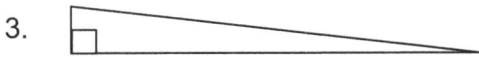

3.   scalene , right

4.   isosceles , acute

5.   equilateral , acute

Scratch Paper

Naming Triangles

Name each triangle by its angles and sides.

1. 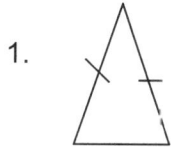   isosceles, acute

2. 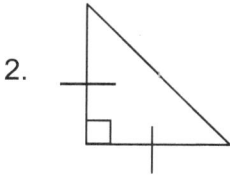   isosceles, right

3.    scalene, right

4.    isosceles obtuse

5. 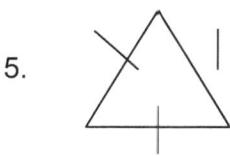   equilateral, acute

Name: _____

Tricky Triangles Recording Sheet

Using the pipe cleaners, try to construct triangles with the given side lengths. If a triangle with the given side lengths cannot be made, answer "no."

Side Lengths	Can a triangle be made? Yes or No
5 cm, 5 cm, 8 cm	
5 in., 8 in., 8 in.	
8 cm, 5 cm, 15 cm	
5 in., 6 in., 10 in.	
11 cm, 12 cm, 14 cm	
8 in., 8 in., 8 in.	

Based on your observations, what must be true in order to construct a triangle?

Scratch Paper

Name: ANSWER KEY

Tricky Triangles Recording Sheet

Using the pipe cleaners, try to construct triangles with the given side lengths. If a triangle with the given side lengths cannot be made, answer "no."

Side Lengths	Can a triangle be made? Yes or No
5 cm, 5 cm, 8 cm	yes
5 in., 8 in., 8 in.	yes
8 cm, 5 cm, 15 cm	no
5 in., 6 in., 10 in.	yes
11 cm, 12 cm, 14 cm	no
8 in., 8 in., 8 in.	yes

Based on your observations, what must be true in order to construct a triangle?

The triangle inequality theorem states that the sum of the lengths of any two sides of a triangle is at most greater than the length of the third side of that triangle.

Name: _____

Warm-up

Find the measure of the missing angle in each triangle below. Do not use a protractor.
Remember: The measures of the three angles in a triangle add up to 180°.

1.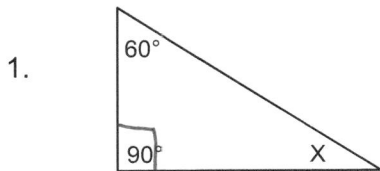

   Measure of ∠X = _30°_

2.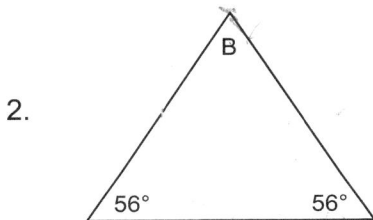

   Measure of ∠B = _56°_ 68

3.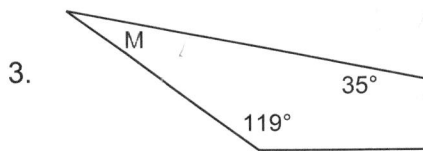

   Measure of ∠M = _26°_   154

4.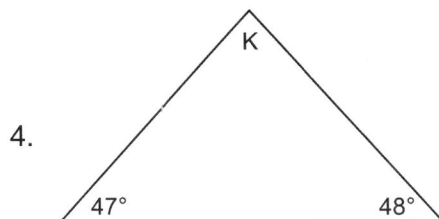

   Measure of ∠K = _85°_

   47+48 = 95

Scratch Paper

Name: ANSWER KEY

Warm-up

Find the measure of the missing angle in each triangle below. Do not use a protractor.
Remember: The measures of the three angles in a triangle add up to 180°.

1. 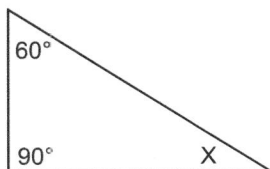   Measure of ∠X = <u>30°</u>

2. 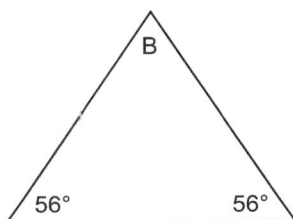   Measure of ∠B = <u>68°</u>

3. 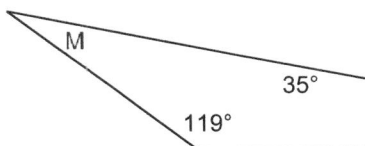   Measure of ∠M = <u>26°</u>

4. 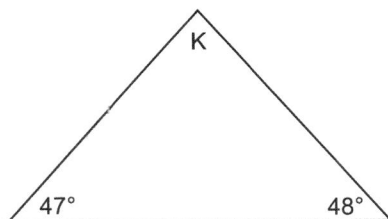   Measure of ∠K = <u>85°</u>

Triangle Vocabulary

Polygon

A closed plane geometric figure in which all the sides are line segments.

Triangle

A three-sided polygon.
The sum of the three angles of a triangle equals 180°.

Acute triangle

A triangle with three acute angles.

Obtuse triangle

A triangle with one obtuse angle.

Right triangle

A triangle with one right angle.

Remember: A triangle is classified by the *largest* of the three angles that form the triangle.

Name: _____

Triangle Classification

Measure each of the angles in each triangle using a protractor. Extend the sides if necessary. Make sure that the measures of the three angles add up to 180°. Record your answers in the Triangle Classification Table. Once you have figured out the measurement of each angle in the triangle, classify the triangle as acute, obtuse, or right.

1.

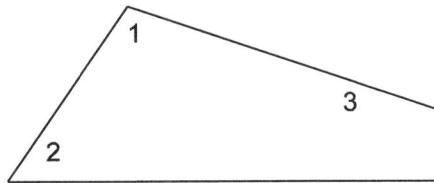

2.

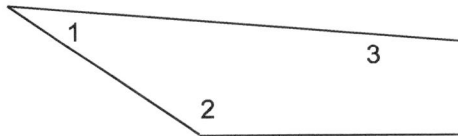

3.

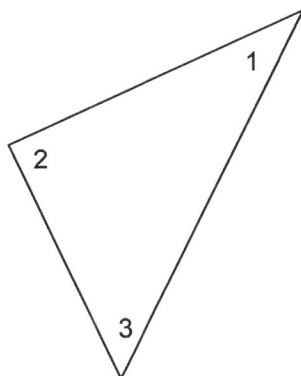

13

Scratch Paper

Triangle Classification

4.

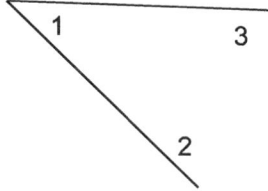

5.

6.

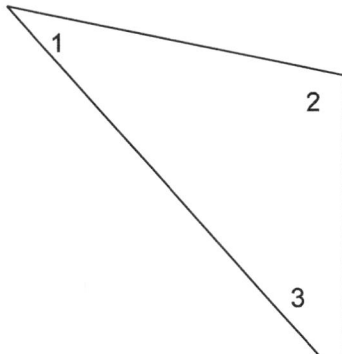

Triangle Classification Table

Triangle Number	Angle 1	Angle 2	Angle 3	Sum of the three angles	Classification
1					
2					
3					
4					
5					
6					

Scratch Paper

Name: <u>ANSWER KEY</u>

Triangle Classification Table

Triangle Number	Angle 1	Angle 2	Angle 3	Sum of the three angles	Classification
1	<u>103°</u>	<u>49°</u>	<u>28°</u>	<u>180°</u>	<u>obtuse</u>
2	<u>30°</u>	<u>125°</u>	<u>25°</u>	<u>180°</u>	<u>obtuse</u>
3	<u>45°</u>	<u>90°</u>	<u>45°</u>	<u>180°</u>	<u>right</u>
4	<u>45°</u>	<u>65°</u>	<u>70°</u>	<u>180°</u>	<u>acute</u>
5	<u>75°</u>	<u>85°</u>	<u>25°</u>	<u>180°</u>	<u>acute</u>
6	<u>37°</u>	<u>97°</u>	<u>46°</u>	<u>180°</u>	<u>obtuse</u>

Reflection

1. To the right are two practice SOL questions. Circle your answers.

The largest angle in △ *ABC* measures 104°. What kind of triangle is △ *ABC*?

A   Equiangular
B   Obtuse
C   Right
D   Acute

2. Explain why you chose that answer to the first question.

_____

_____

_____

_____

The angles in △ *ABC* measure 27°, 73°, and 80°. What kind of triangle is △ *ABC*?

F   Equiangular
G   Acute
H   Obtuse
J   Right

3. Explain why you chose that answer to the second question.

_____

_____

_____

4. Can a triangle have more than one right angle?_____Explain.

_____

_____

_____

Scratch Paper

Name: ANSWER KEY

Reflection

1. To the right are two practice SOL questions. Circle your answers.

   <u>B, G</u>

2. Explain why you chose that answer to the first question.

   <u>A triangle is classified by its largest angle. A 104° angle is an obtuse angle. Therefore, the triangle is an obtuse triangle.</u>

3. Explain why you chose that answer to the second question.

   <u>All the angles in the triangle are acute. Therefore, the triangle is an acute triangle.</u>

4. Can a triangle have more than one right angle? <u>No</u> Explain.

   <u>Two right angles add up to be 180°, and because the *three* angles of a triangle must add up to be 180°, there could not be another angle.</u>

The largest angle in $\triangle ABC$ measures 104°. What kind of triangle is $\triangle ABC$?

A Equiangular
(B) Obtuse
C Right
D Acute

The angles in $\triangle ABC$ measure 27°, 73°, and 80°. What kind of triangle is $\triangle ABC$?

F Equiangular
(G) Acute
H Obtuse
J Right

Scratch Paper

Quadrilateral Concept Card

Polygons

These figures are polygons:

These figures are <u>not</u> polygons:

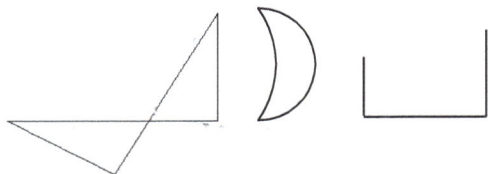

Which of these figures are polygons? (circle)

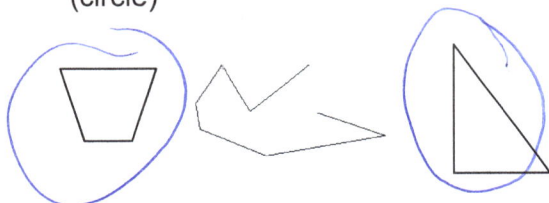

Draw your own example of a polygon.

Draw your own example of a non-polygon.

What is a polygon?

A polygon is   A Plane fishure

_____.

Quadrilaterals

These figures are quadrilaterals:

These figures are <u>not</u> quadrilaterals:

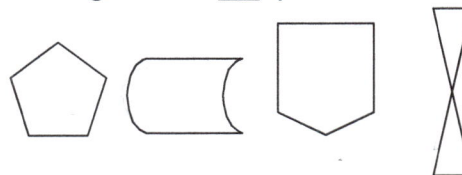

Which of these figures are quadrilaterals? (circle)

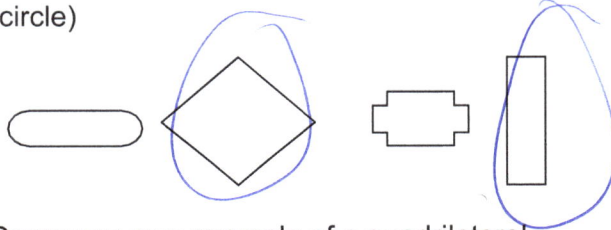

Draw your own example of a quadrilateral.

Draw your own example of a non-quadrilateral.

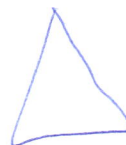

What is a quadrilateral?

A quadrilateral is   A shape with

4 angles and corners.

Quadrilateral Concept Card

Polygons

These figures are polygons:

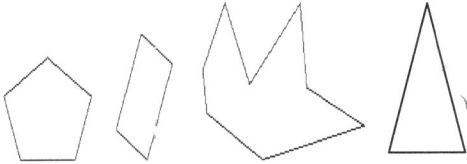

These figures are <u>not</u> polygons:

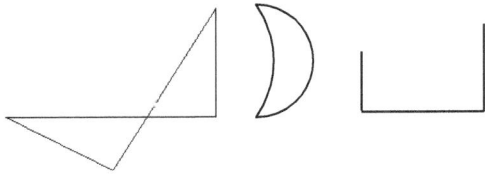

Which of these figures are polygons? (circle)

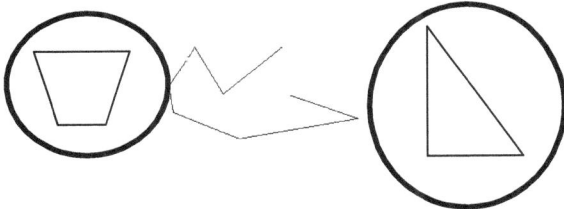

Draw your own example of a polygon.

<u>Drawing will vary.</u>

Draw your own example of a non-polygon.

<u>Drawing will vary. Sample answer:</u>

What is a polygon? <u>Sample answer:</u>

<u>A polygon is a simple, closed, plane figure formed by three or more straight lines.</u>

Quadrilaterals

These figures are quadrilaterals:

These figures are <u>not</u> quadrilaterals:

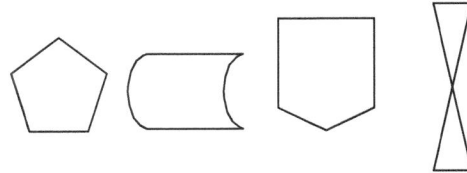

Which of these figures are quadrilaterals? (circle)

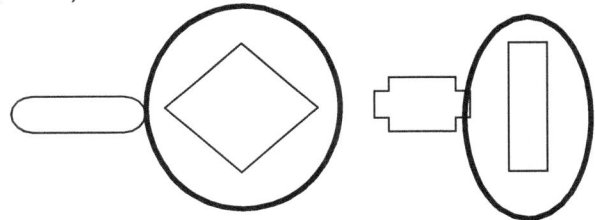

Draw your own example of a quadrilateral.

<u>Drawing will vary.</u>

Draw your own example of a non-quadrilateral.

<u>Drawing will vary. Sample answer:</u>

What is a quadrilateral? <u>Sample answer:</u>

<u>A quadrilateral is a four-sided polygon.</u>

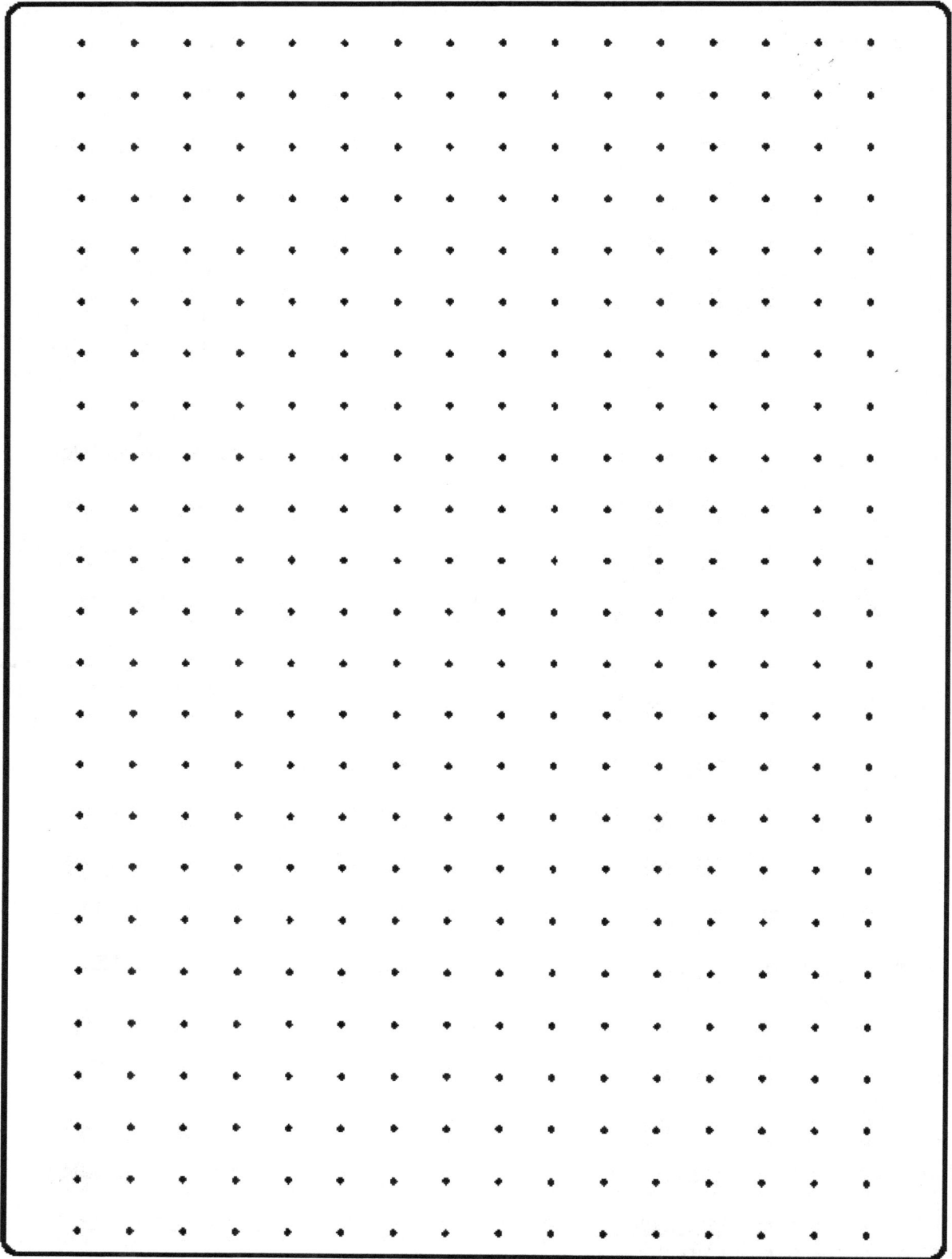

Dot Paper

Quadrilateral Study Guide

Fill in the blanks, and draw the figures as directed.

1. A _quadrilateral_ is a polygon with four sides. Draw several examples of this below.

2. A _parallelogram_ is a quadrilateral in which both pairs of opposite sides are parallel.

3. Properties of a parallelogram include the following:

    a. A diagonal divides a parallelogram into two congruent _triangles_ .

    b. The opposite sides of a parallelogram are _parall_ .

    c. The opposite angles of a parallelogram are _congruent_ .

For questions 4–8, refer to the drawings on the right.

4. A _rectangle_ is a parallelogram with four right angles. Since a _rectangle_ is a parallelogram, it has the same properties as those of a parallelogram.

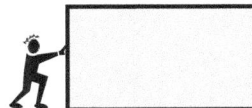

5. A _square_ is a rectangle with four congruent sides. Since a _square_ is a rectangle, it has all the properties of a rectangle and of a parallelogram.

6. A _rhombus_ is a parallelogram with four congruent sides. Opposite angles of a _rhombus_ are congruent. Since a _rhombus_ is a parallelogram, it has all the properties of a parallelogram.

7. A _trapezoid_ is a quadrilateral with exactly one pair of parallel sides.

Quadrilateral Study Guide

Fill in the blanks, and draw the figures as directed.

1. A <u>quadrilateral</u> is a polygon with four sides. Draw several examples of this below.

2. A <u>parallelogram</u> is a quadrilateral in which both pairs of opposite sides are parallel.

3. Properties of a parallelogram include the following:

    a. A diagonal divides a parallelogram into two congruent <u>triangles</u>.

    b. The opposite sides of a parallelogram are <u>parallel/congruent</u>.

    c. The opposite angles of a parallelogram are <u>congruent</u>. For

questions 4–8, refer to the drawings on the right.

4. A <u>rectangle</u> is a parallelogram with four right angles. Since a _
    <u>rectangle</u> is a parallelogram, it has the same properties as those of a
    parallelogram. (Discuss with students that *square* also fits this
    definition, since a square is a special type of rectangle.)

5. A <u>square</u> is a rectangle with four congruent sides. Since a <u>square</u> is a rectangle,
    it has all the properties of a rectangle and of a parallelogram.

6. A <u>rhombus</u> is a parallelogram with four congruent sides. Opposite
    angles of a <u>rhombus</u> are congruent. Since a <u>rhombus</u> is a
    parallelogram, it has all the properties of a parallelogram. (Discuss
    with students that a square is a special rhombus, so it also fits this definition, but a square
    must have four right angles.)

7. A <u>trapezoid</u> is a quadrilateral with exactly one pair of parallel sides.

Scratch Paper

Tangram Template

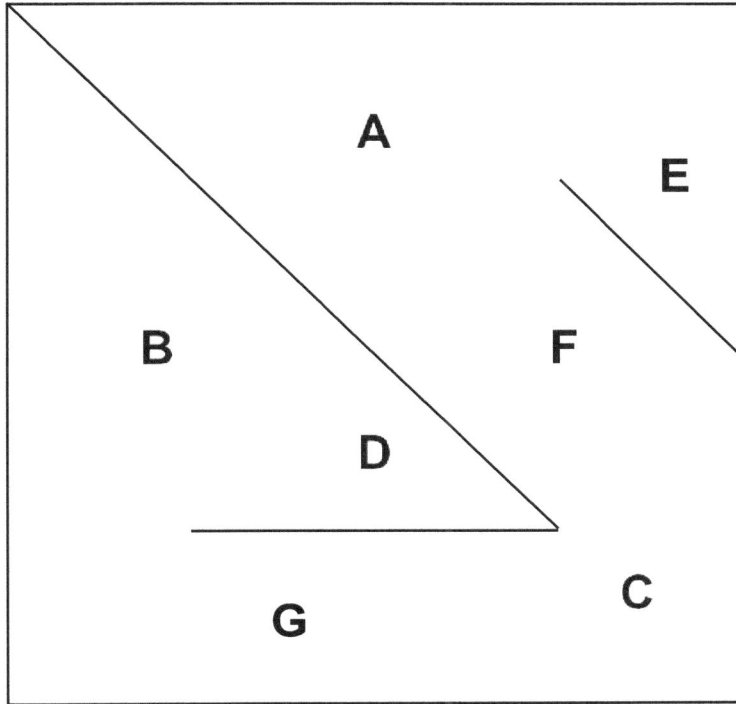

A

E

B

F

D

C

G

Name: _____

Tangram Activity Chart

Use the number of tangram pieces specified in the first column to form each of the geometric figures shown in the top row. As you make each shape, list in the proper box the pieces you use to make it. *Some problems may have more than one solution, while others may have no solution.*

Number of pieces	Square	Rectangle	Triangle	Trapezoid	Parallel-ogram
2					
3					
4					
5					
6					
7					

Scratch Paper

Tangram Activity Chart

Use the number of tangram pieces specified in the first column to form each of the geometric figures shown in the top row. As you make each shape, list in the proper box the pieces you use to make it. *Some problems may have more than one solution, while others may have no solution.*

<u>(Teacher's Note: For each problem with multiple solutions, only one is included here.)</u>

Number of pieces	Square	Rectangle	Triangle	Trapezoid	Parallel-ogram
2	<u>D & E</u>	<u>Not possible</u>	<u>D & E</u>	<u>D & F</u>	<u>D & E</u>
3	<u>D, E, & C</u>	<u>D, C, & E</u>	<u>D, E, & C</u>	<u>A, B, & C</u>	<u>D, E, & C</u>
4	<u>D, E, C, & B</u>	<u>D, E, C, & B</u>	<u>D, E, C, & B</u>	<u>D, E, A, & B</u>	<u>D, E, C, & B</u>
5	<u>D, E, F, G, & C</u>	<u>D, E, F, G, & C</u>	<u>D, E, F, G, & C</u>	<u>D, E, F, G, & C</u>	<u>D, E, F, G, & C</u>
6	<u>Not possible</u>	<u>C, E, F, D, G, & A</u>	<u>Not possible</u>	<u>C, E, F, D, G, & A</u>	<u>C, E, F, D, G, & A</u>
7	<u>All pieces</u>	<u>All pieces</u>	<u>All pieces</u>	<u>All pieces</u>	<u>All pieces</u>

Tangram Puzzles

Can you make these figures, using all seven tangram pieces? Make a sketch of your solutions.

1.

2.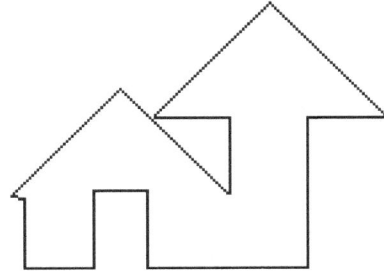

3.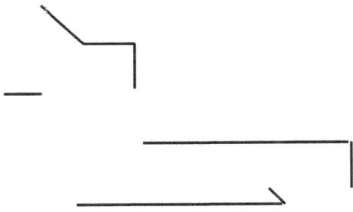

4.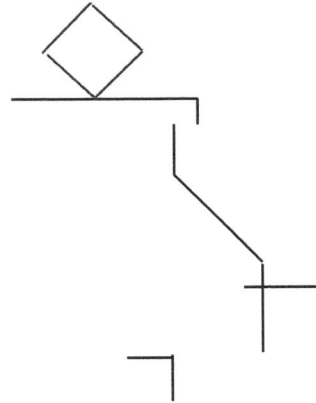

5.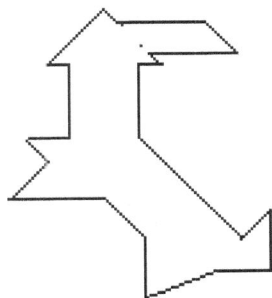

6.

Name: _____

Assessment Questions

1. What strategy did you use to find the shapes?

2. Is there a basic shape that could be used to make all the figures?

3. Which pieces did you tend to use more than others? Why?

4. What are some other ways to construct each of the shapes?

5. Why is there no six-piece square?

Similar and Congruent Triangle Sorting Pieces

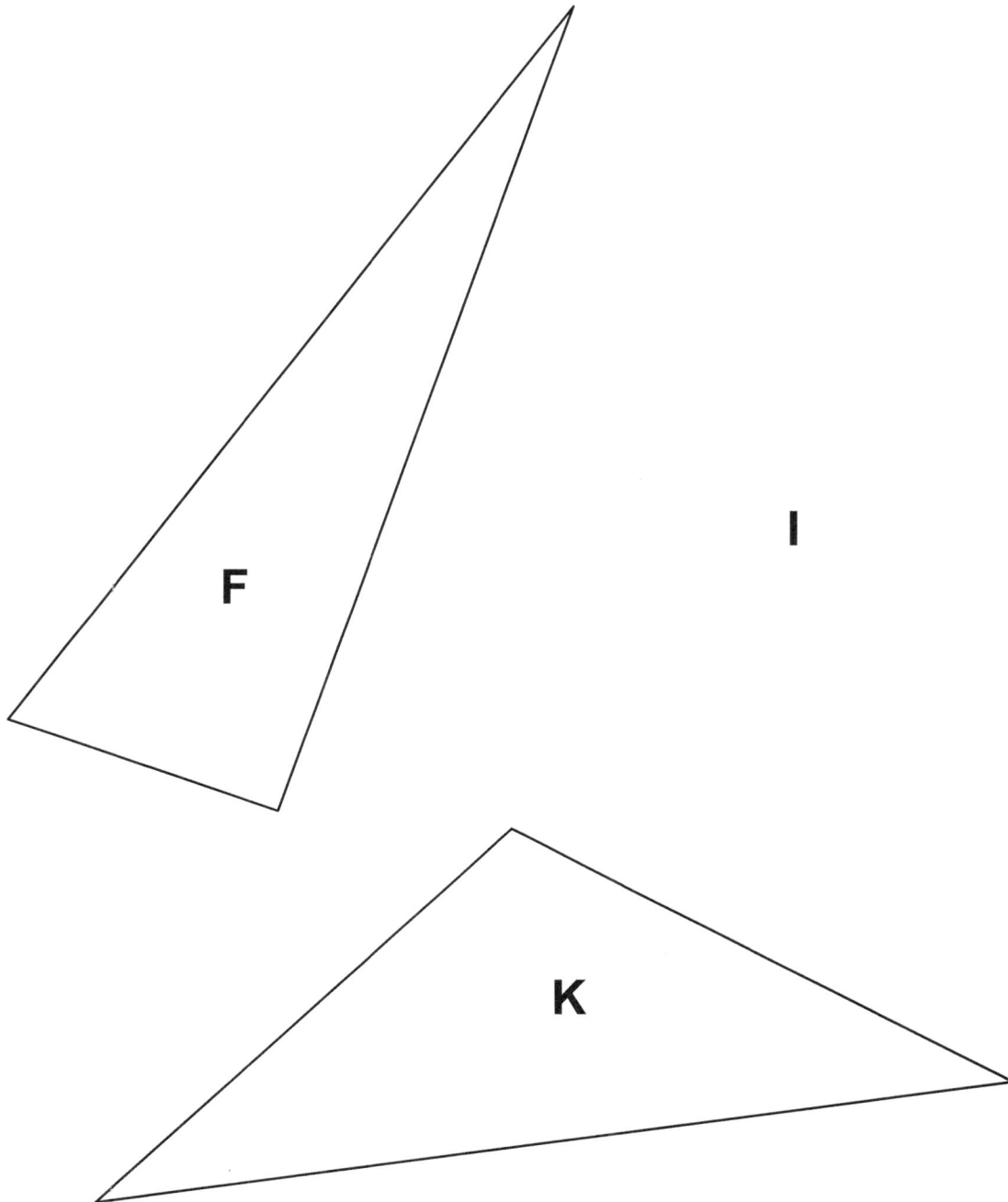

F

I

K

Similar and Congruent Triangle Sorting Pieces

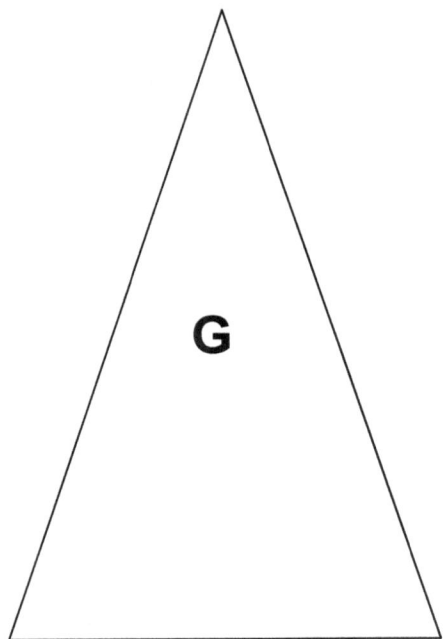

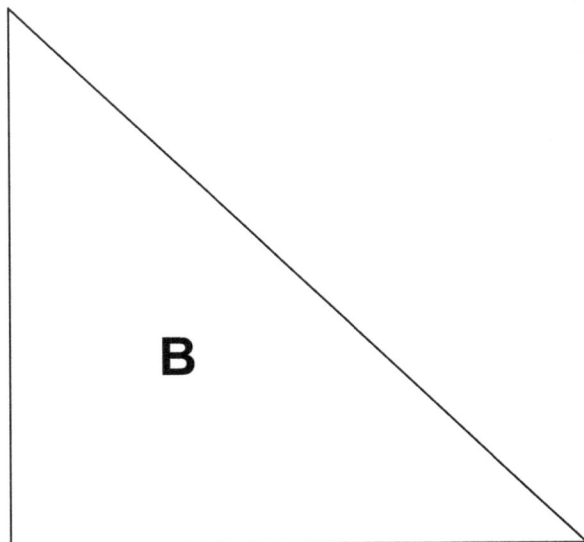

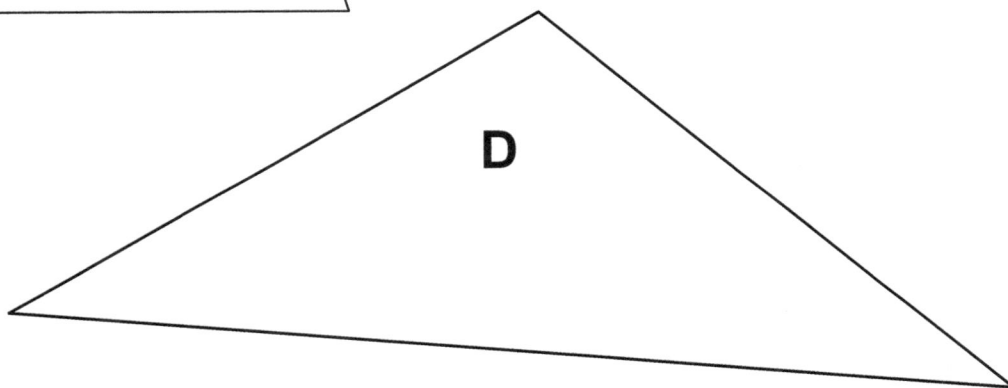

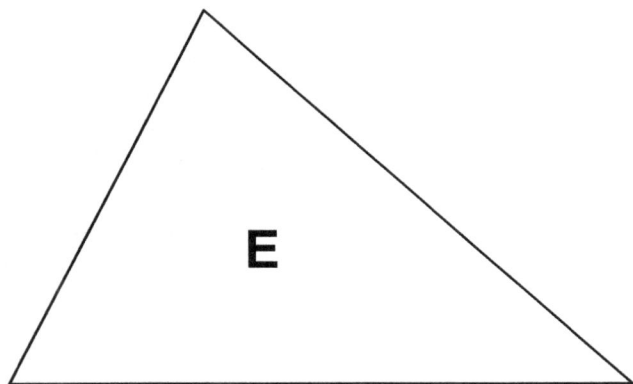

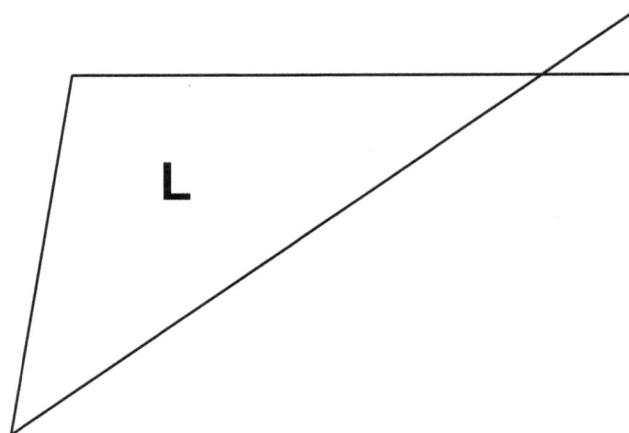

G

B

D

E

L

Similar and Congruent Triangle Sorting Pieces

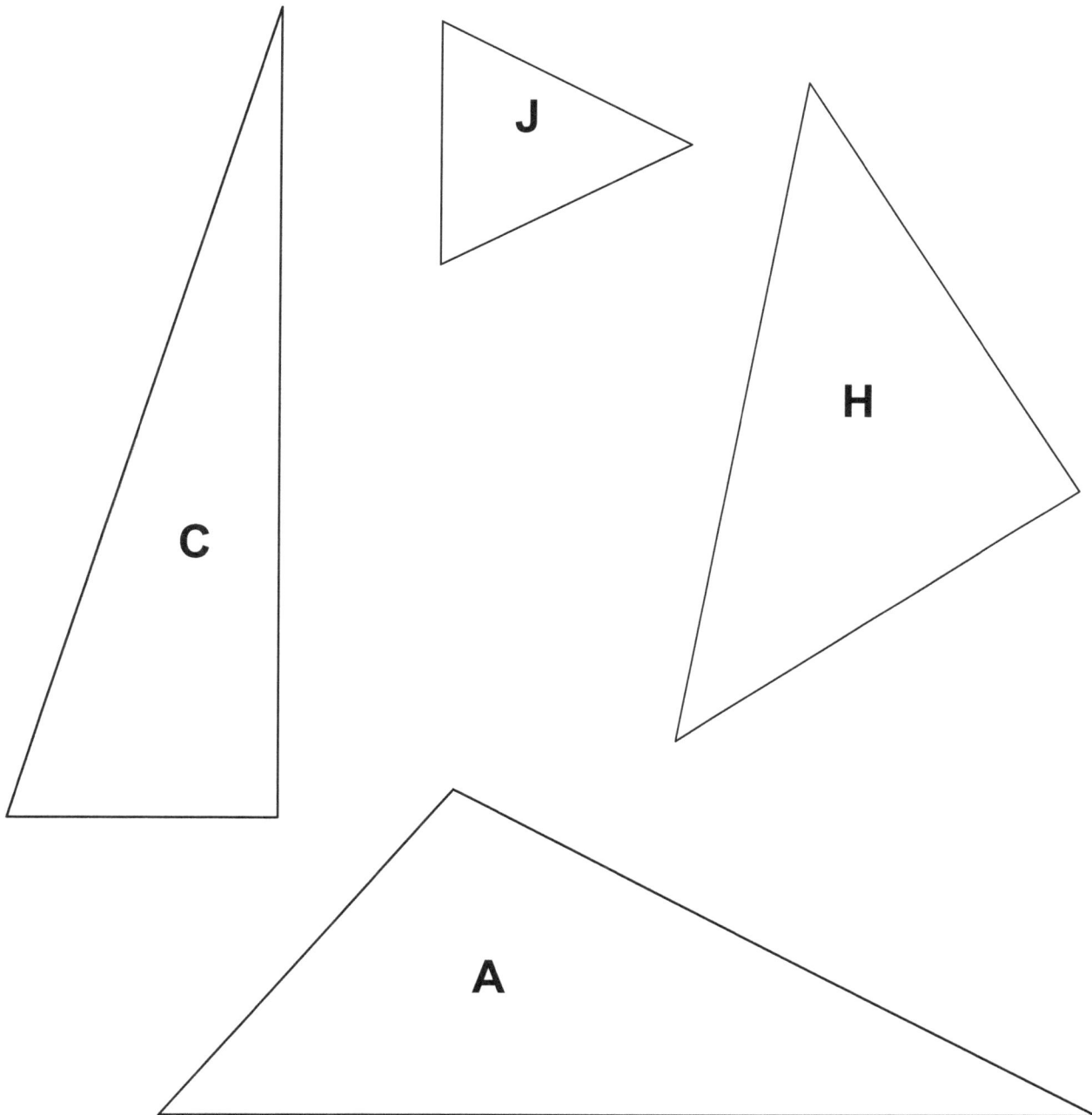

C

J

H

A

Applying the Lesson

In the first box in each row, draw a simple figure. In the second box in that row, draw a figure <u>similar</u> to the original. In the third box, draw a figure <u>congruent</u> to the original figure.

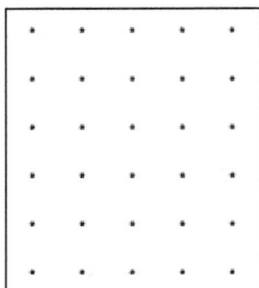

Original Figure   **Similar Figure**   **Congruent Figure**

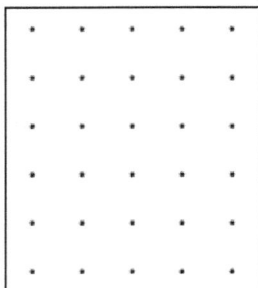

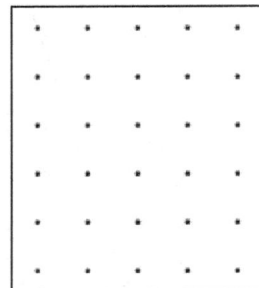

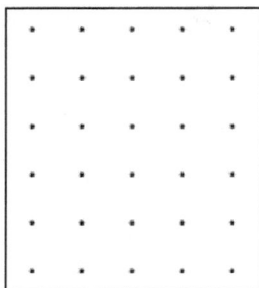

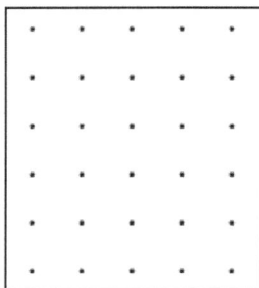

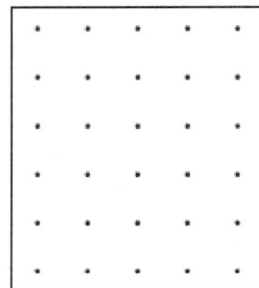

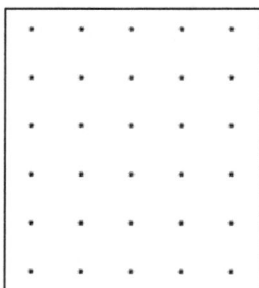

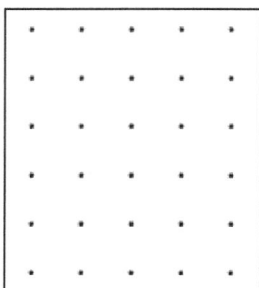

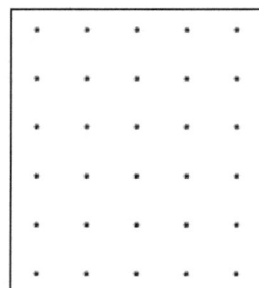

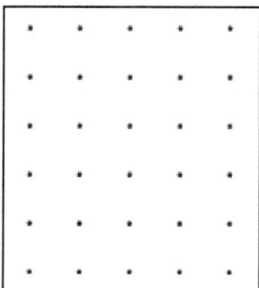

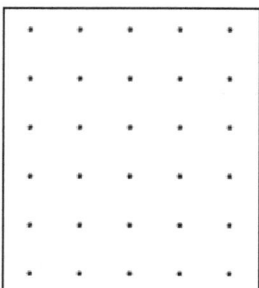

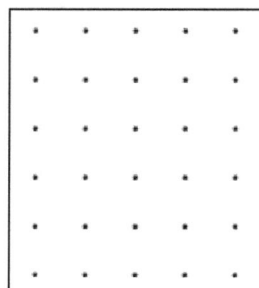

Scratch Paper

Applying the Lesson

In the first box in each row, draw a simple figure. In the second box in that row, draw a figure <u>similar</u> to the original. In the third box, draw a figure <u>congruent</u> to the original figure.

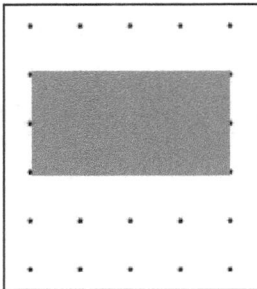

Original Figure   **Similar Figure**   **Congruent Figure**

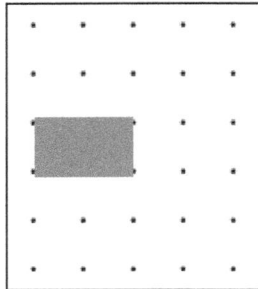

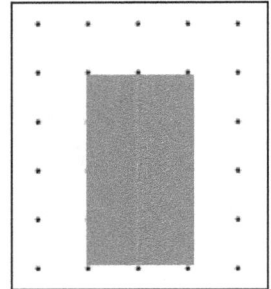

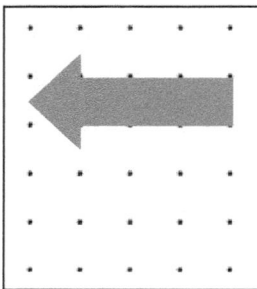

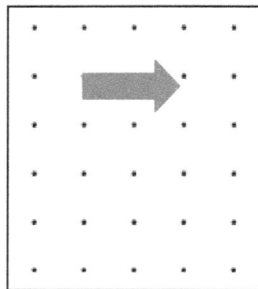

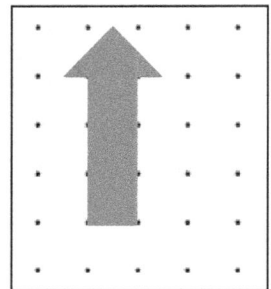

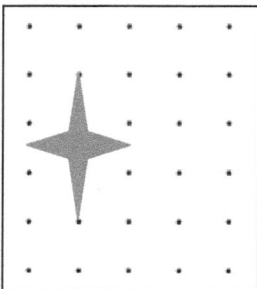

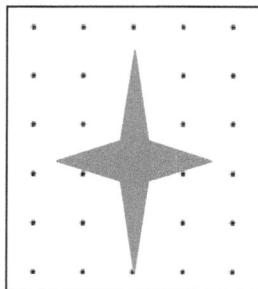

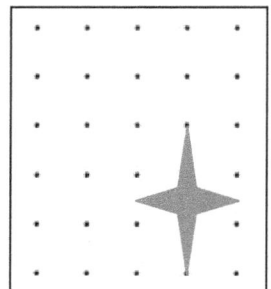

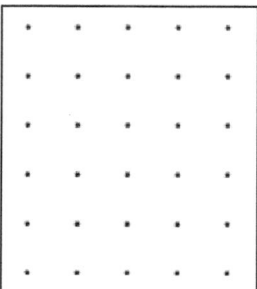

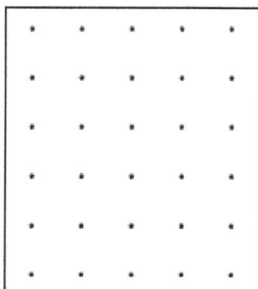

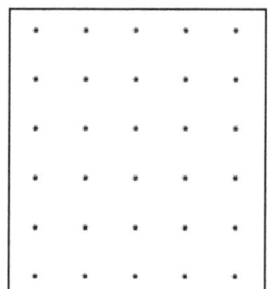

Name: _____

Similar or Congruent?

Label each pair of figures below as **similar**, **congruent**, or **neither**. Then, justify your answer with its mathematical definition.

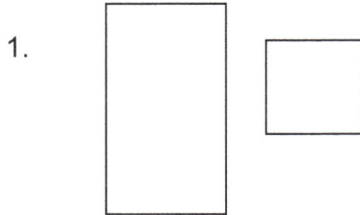

1. These figures are ___similar___ because Here the same thing

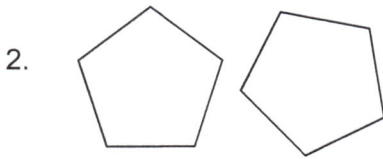

2. These figures are ___similar___ because Here just turned around.

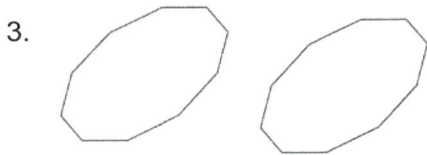

3. These figures are ___similar___ because its the same

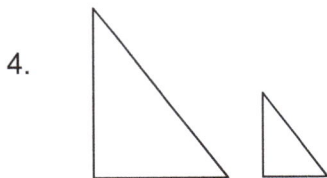

4. These figures are ___similar___ because It just smaller

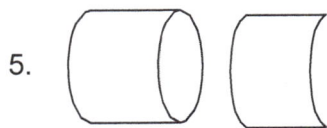

5. These figures are ___congruent___ because its not the same.

Scratch Paper

Name: ANSWER KEY

Similar or Congruent?

Label each pair of figures below as **similar, congruent,** or **neither**. Then, justify your answer with its mathematical definition.

1.

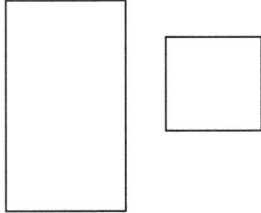

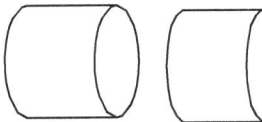

These figures are <u>neither</u> because

<u>they are neither the same shape nor the same size.</u>

_____

2.

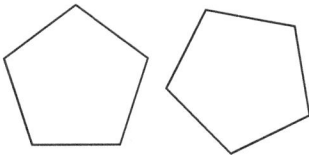

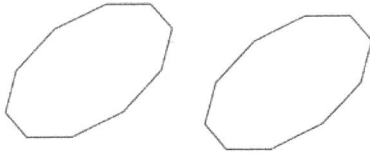

These figures are <u>congruent</u> because

<u>they are exactly the same in shape and size.</u>

_____

3.

These figures are <u>congruent</u> because

<u>they are exactly the same in shape and size.</u>

_____

4.

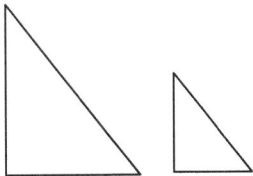

These figures are <u>similar</u> because

<u>they are exactly the same in shape but different in size.</u>

_____

5.

These figures are <u>neither</u> because

<u>they are neither the same shape nor the same size.</u>

_____

45

Name: _____

Similar and Congruent Figures Concept Card

Similar Figures	Congruent Figures

Similar Figures

These pairs of figures are <u>similar</u>.

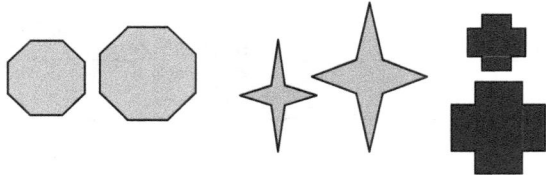

Congruent Figures

These pairs of figures are <u>congruent</u>.

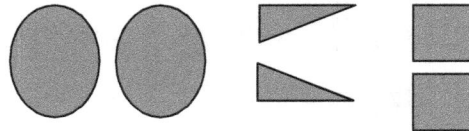

These pairs of figures are <u>nonsimilar</u>.

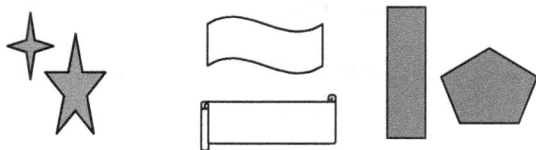

These pairs of figures are <u>noncongruent</u>.

Which of these pairs of figures are similar?

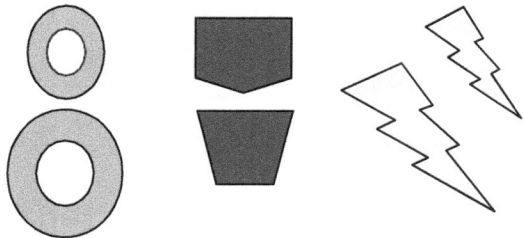

Which of these pairs of figures are congruent?

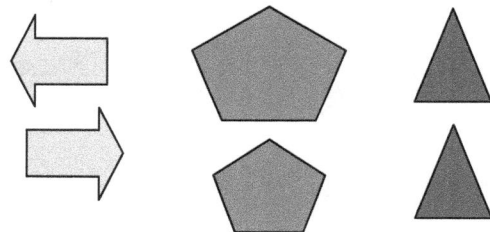

Draw your own pair of similar figures.

Draw your own pair of congruent figures.

Draw your own pair of nonsimilar figures.

Draw your own pair of noncongruent figures.

What are similar figures?

<u>Similar figures are</u> _____

_____ .

What are congruent figures?

<u>Congruent figures are</u> _____

_____ .

Scratch Paper

Similar and Congruent Figures Concept Card

| Similar Figures | Congruent Figures |

Similar Figures

These pairs of figures are <u>similar</u>.

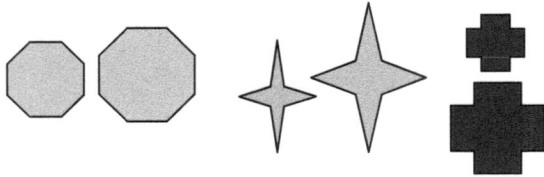

These pairs of figures are <u>nonsimilar</u>.

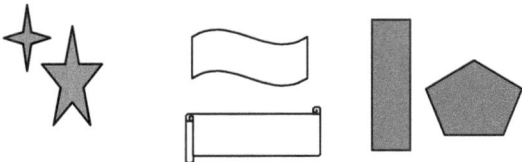

Which of these pairs of figures are similar?

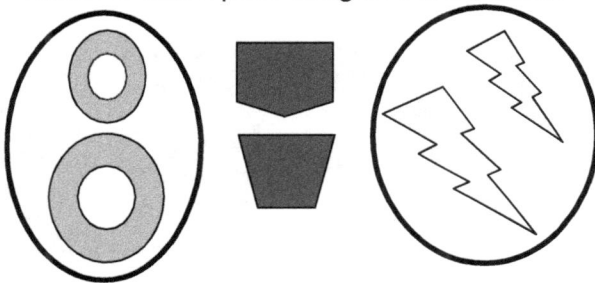

Draw your own pair of similar figures.
<u>Sample answer:</u>

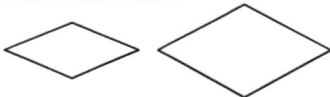

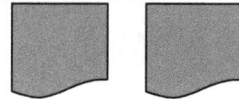

Draw your own pair of nonsimilar figures.
<u>Sample answer:</u>

What are similar figures? <u>Sample answer:</u>

<u>Similar figures are two figures that are exactly the same in shape but not necessarily the same in size.</u>

Congruent Figures

These pairs of figures are <u>congruent</u>.

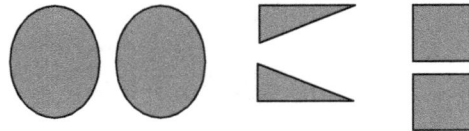

These pairs of figures are <u>noncongruent</u>.

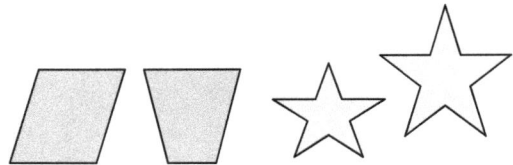

Which of these pairs of figures are congruent?

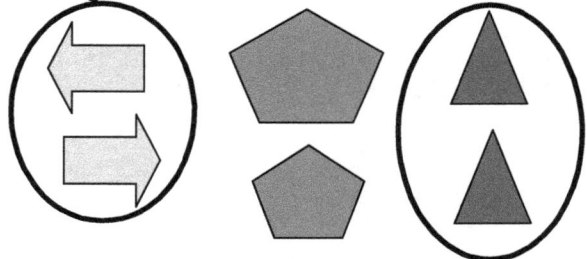

Draw your own pair of congruent figures.
<u>Sample answer:</u>

Draw your own pair of noncongruent figures.
<u>Sample answer:</u>

What are congruent figures? <u>Sample answer:</u>
<u>Congruent figures are two figures that are exactly same in shape and size.</u>

48

Similar and Congruent Figures Sorting Pieces

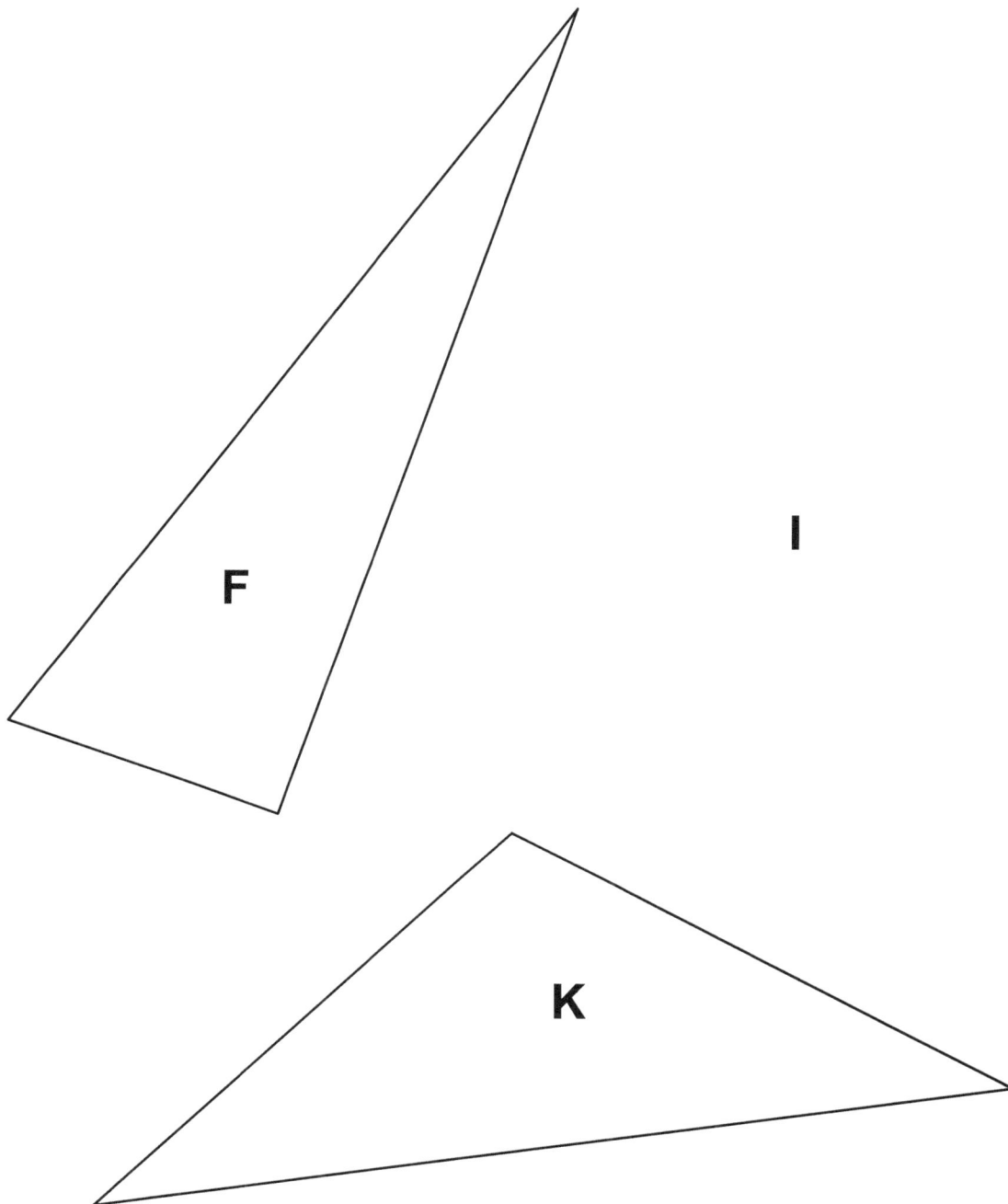

F

I

K

Similar and Congruent Figures Sorting Pieces

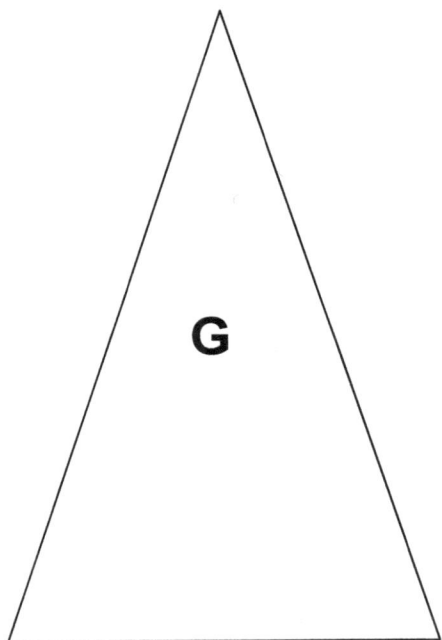

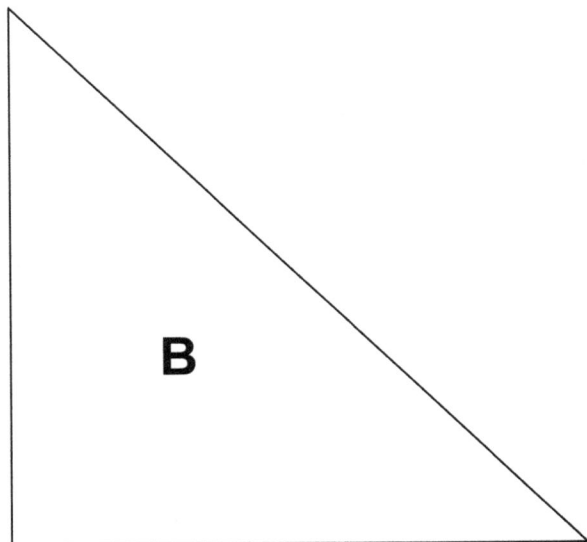

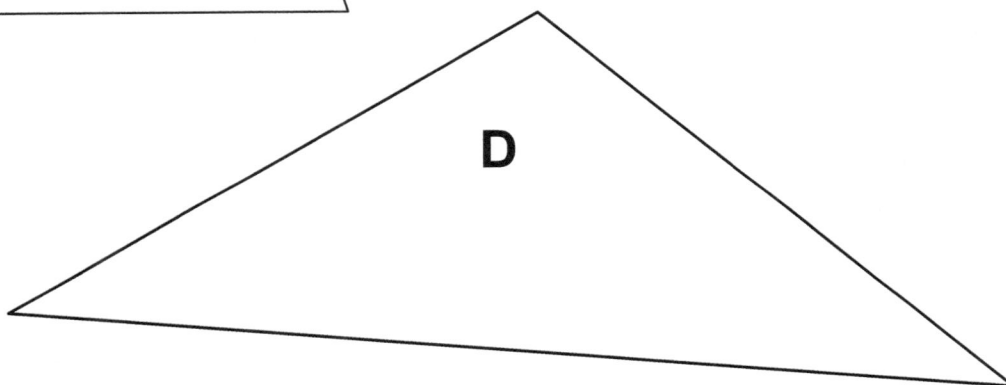

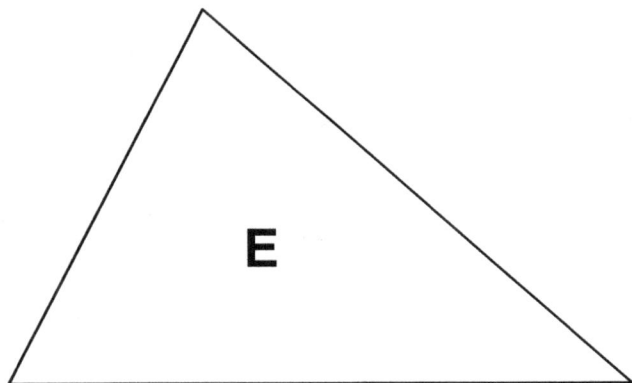

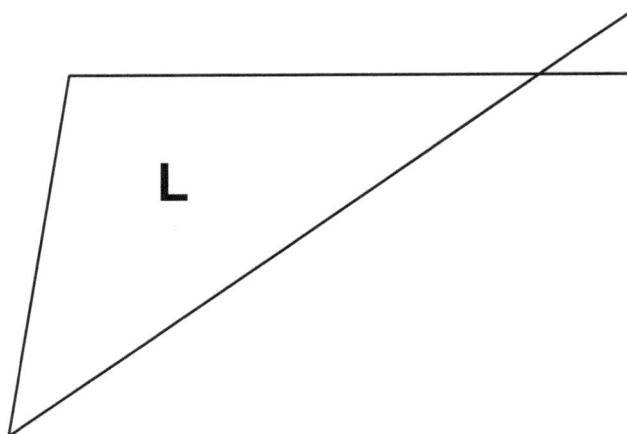

G

B

D

E

L

Similar and Congruent Figures Sorting Pieces

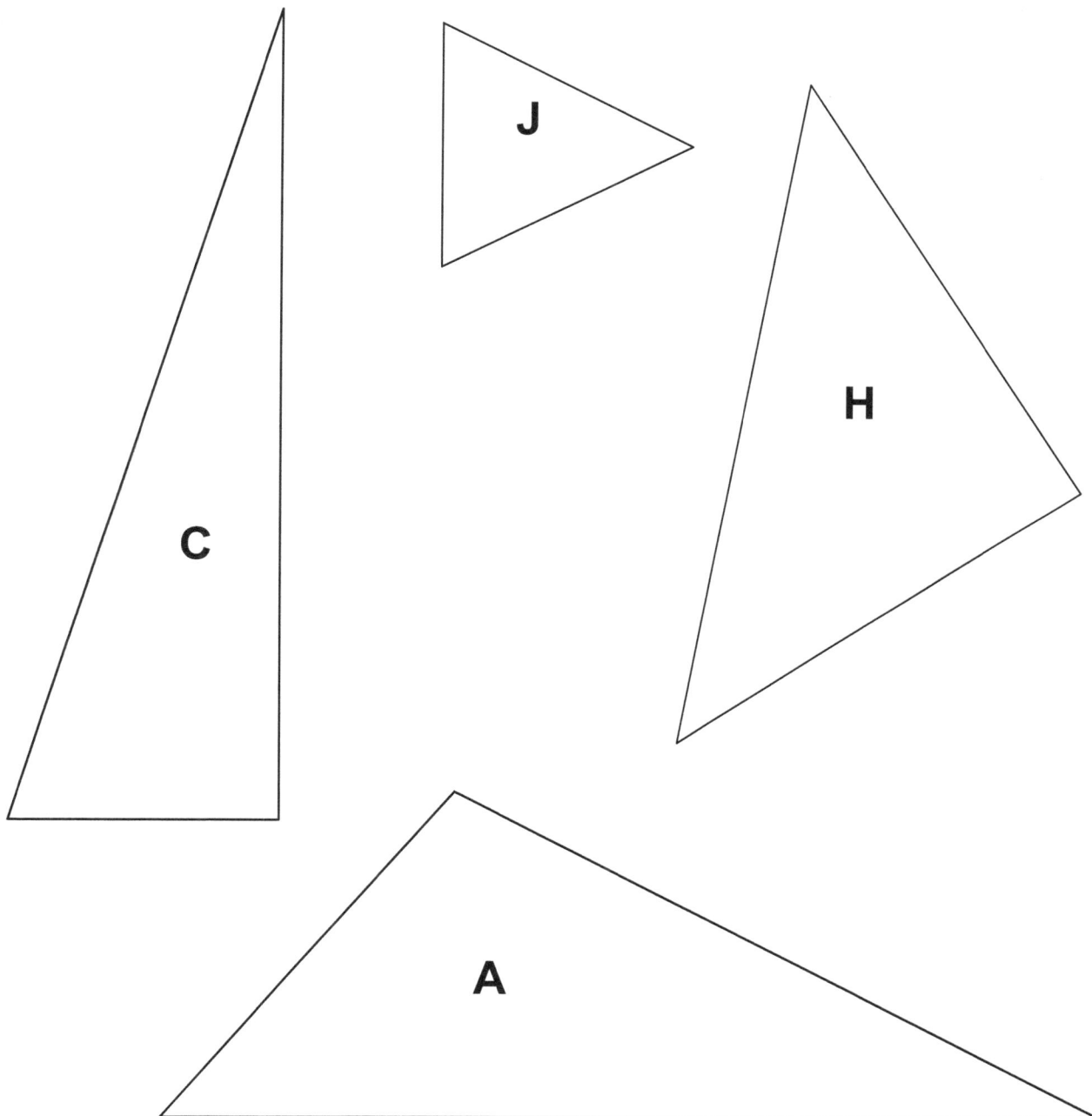

J

C

H

A

Similar and Congruent Figures Sorting Pieces

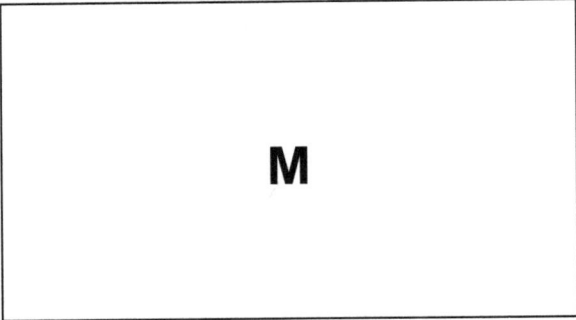

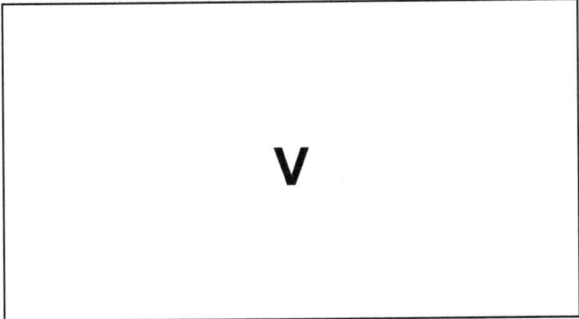

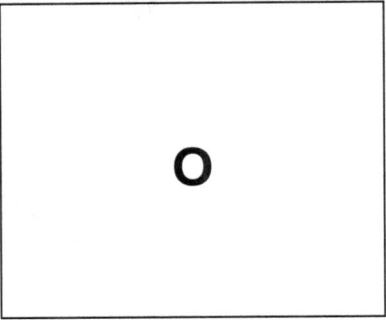

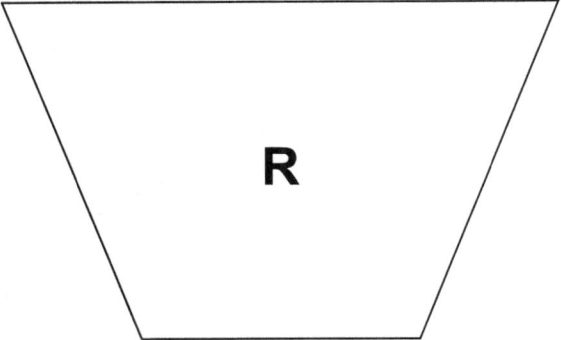

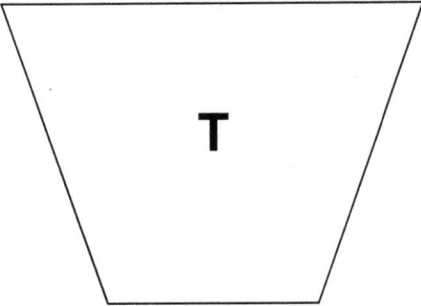

M

V

O

S

T

R

Similar and Congruent Figures Sorting Pieces

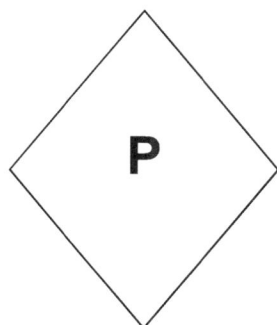

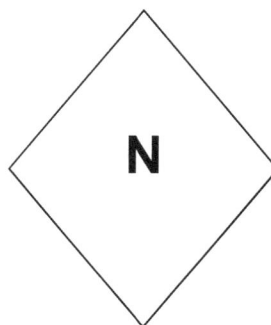

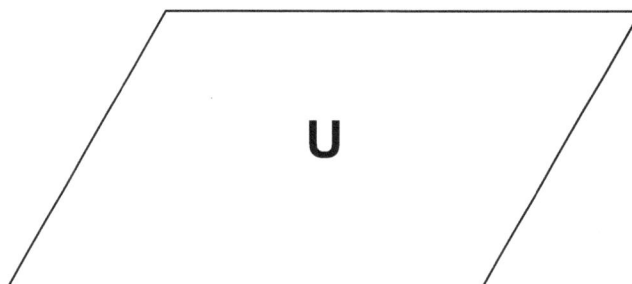

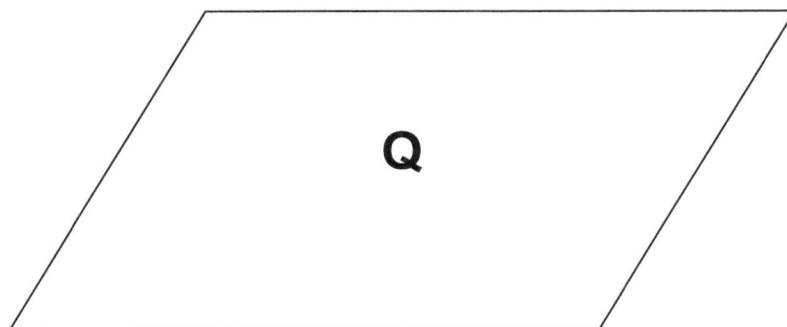

P

N

U

Q

Similar or Congruent Figures Chart

Sort the figures into pairs according to whether they are similar or congruent. Then, identify the pairs by writing their letters in the chart below.

SIMILAR FIGURES	CONGRUENT FIGURES

Scratch Paper

Similar or Congruent Figures Chart

Sort the figures into pairs according to whether they are similar or congruent. Then, identify the pairs by writing their letters in the chart below.

SIMILAR FIGURES	CONGRUENT FIGURES
O & S	P & N
R & T	M & V
U & Q	C & F
J & G	H & B
A & E	I & L
	D & K

Name: _____

Similar or Congruent?

Determine whether each pair of figures below is **similar** or **congruent**, and give a brief explanation of your answer.

1.
These figures are_____because

_____

_____.

2.
These figures are_____because

_____

_____.

3.
These figures are_____because

_____

_____.

4. Draw a pair of quadrilaterals that are similar and another pair that are congruent. Label the pairs.

Answer *True* or *False* to the following questions. If false, explain why.

5. Congruent figures are always similar. _____

6. Similar figures are never congruent. _____

_____

_____

Scratch Paper

Name: ANSWER KEY

Similar or Congruent?

Determine whether each pair of figures below is **similar** or **congruent**, and give a brief explanation of your answer.

1.    These figures are <u>congruent</u> because

   <u>they are exactly the same in shape and size.</u>

   _____

2. 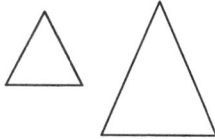   These figures are <u>similar</u> because

   <u>they are exactly the same in shape but different in size.</u>

   _____

3. 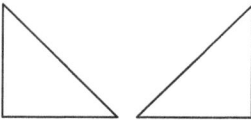   These figures are <u>congruent</u> because

   <u>they are exactly the same in shape and size.</u>

   _____

4. Draw a pair of quadrilaterals that are similar and another pair that are congruent. Label the pairs. <u>Sample answers:</u>

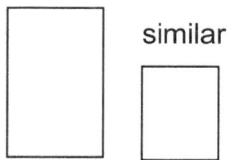 similar

 congruent

Answer *True* or *False* to the following questions. If false, explain why.

5. Congruent figures are always similar. <u>True</u>

6. Similar figures are never congruent. <u>False. They may be the same in size, and if they are, they are congruent.</u>

Name: _____

Congruent Figures Concept Card

These pairs of figures are <u>congruent</u>.

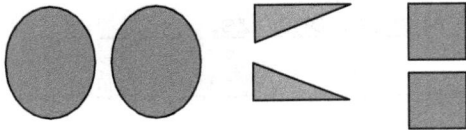

These pairs of figures are <u>noncongruent</u>.

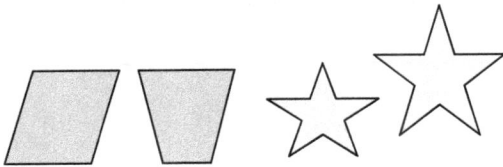

Which of these pairs of figures are congruent?

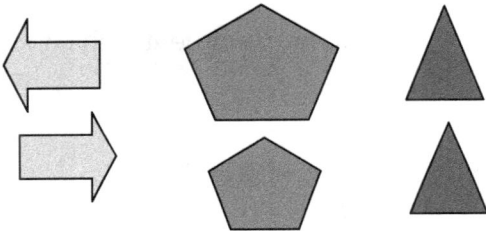

Draw your own pair of congruent figures.

Draw your own pair of noncongruent figures.

What are congruent figures?

<u>Congruent figures are</u> _____

_____.

Scratch Paper

Congruent Figures Concept Card

These pairs of figures are <u>congruent</u>.

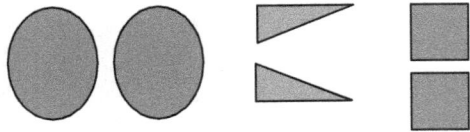

These pairs of figures are <u>noncongruent</u>.

Which of these pairs of figures are congruent?

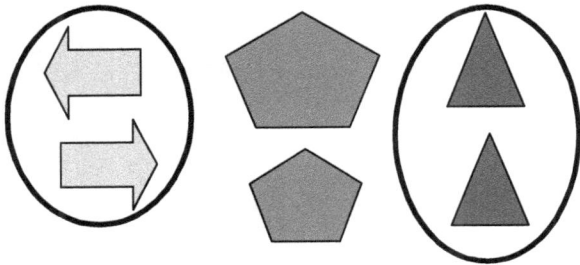

Draw your own pair of congruent figures.
<u>Sample answer:</u>

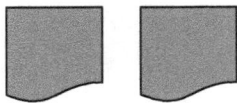

Draw your own pair of noncongruent figures.
<u>Sample answer:</u>

What are congruent figures? <u>Sample answer:</u>

<u>Congruent figures are two figures that are exactly the same in shape and size.</u>

62

Name: _____

Rectangle Templates

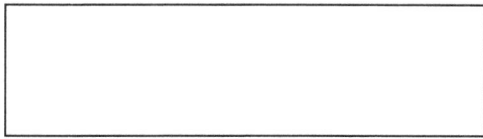

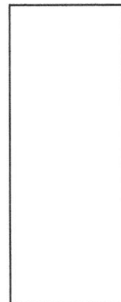

Name: _____

Congruent Figures Worksheet

Look carefully each row of figures. Circle the figures in that row that are congruent to the first figure in the row, and label them "translation," "rotation," or "reflection." Some figures might be labeled with two terms.

1.

2.

3.

4.

5.

6.

7.

8.

Scratch Paper

Congruent Figures Worksheet

Look carefully each row of figures. Circle the figures in that row that are congruent to the first figure in the row, and label them "translation," "rotation," or "reflection." Some figures might be labeled with two terms.

1.

rotation   rotation

2.

rotation or
reflection   rotation

3.

rotation   translation

4.

rotation   rotation

5.

translation

translation

6.

rotation   translation

7.

reflection   reflection
and rotation   rotation

8.

reflection

66

Reflecting on Congruent Figures

In the space below, draw a congruent figure for each figure shown.

1.

2.

3.

4.

5.

6.

Reflecting on Congruent Figures

7. **Which 2 shapes below are congruent?**

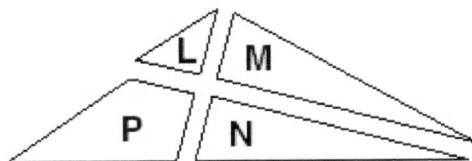

   A L and P
   B M and N
   C N and L
   D P and N

9. **Which figures appear to be congruent?**

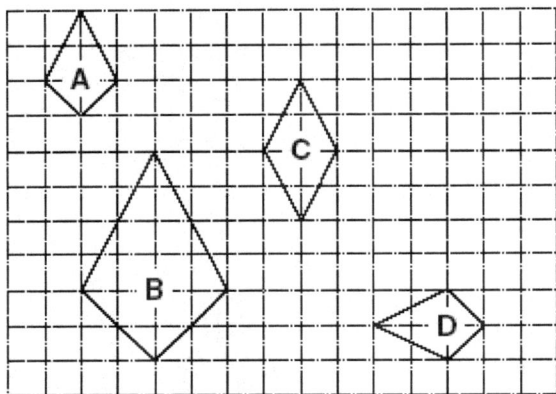

   A   A and B
   B   B and D
   C   C and D
   D   D and A

8. Wayne cut this shape out of a piece of paper.

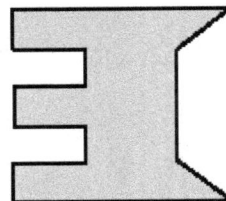

Which of the following is missing a piece exactly the same size and shape as the piece shown above?

F
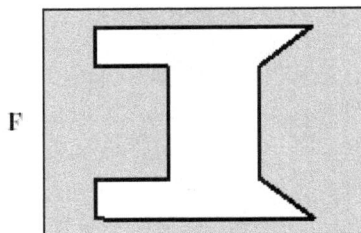

G
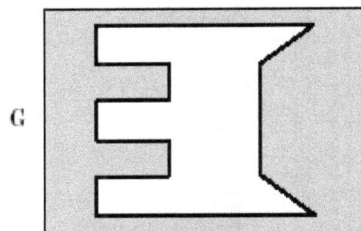

H
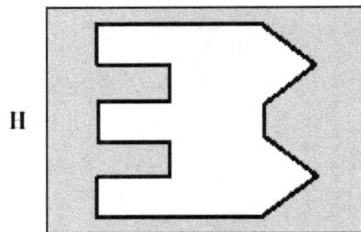

J
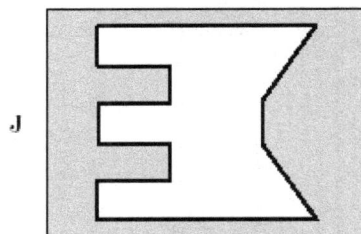

Reflecting on Congruent Figures

In the space below, draw a congruent figure for each figure shown.

1.

2.

3.

4.

5.

6.

Reflecting on Congruent Figures

7. **Which 2 shapes below are congruent?**

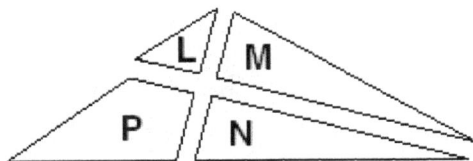

   A L and P
   (B) M and N
   C N and L
   D P and N

8. Wayne cut this shape out of a piece of paper.

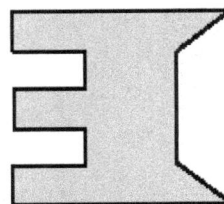

   Which of the following is missing a piece exactly the same size and shape as the piece shown above?

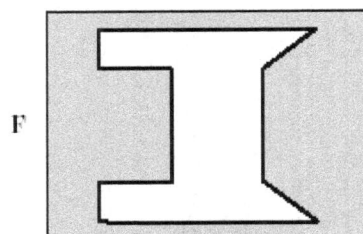

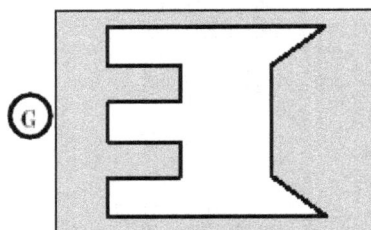

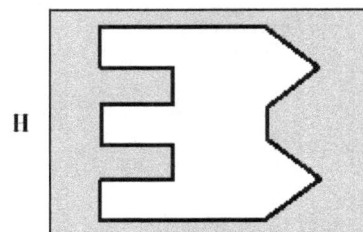

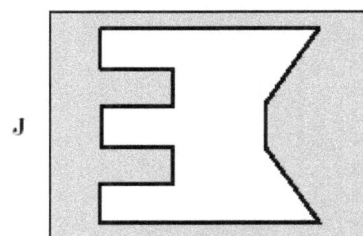

9. **Which figures appear to be congruent?**

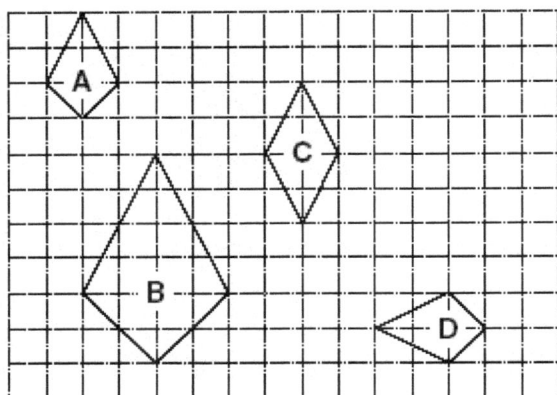

   A A and B
   B B and D
   C C and D
   (D) D and A

F

(G)

H

J

Warm-up

Measure each line segment, using a centimeter ruler, and record each measurement.

1.   _____   Measure = _____
     K   M

2.   _____   Measure = _____
     G   T

3.   _____   Measure = _____
     W   R

4.   _____   Measure = _____
     E   L

Measure each angle in degrees, using a protractor, and record each measure.

5.   Measure = _____

6.   Measure = _____

7.   Measure = _____

8.   Define *congruent*:_____

Scratch Paper

Name: ANSWER KEY

Warm-up

Measure each line segment, using a centimeter ruler, and record each measurement.

1. K _____ M

    Measure = <u>4 1/2 cm or 4.5 cm</u>

2. G _____ T

    Measure = <u>7 9/10 cm or 7.9 cm</u>

3.

    W _____ R

    Measure = <u>2 9/10 cm or 2.9 cm</u>

4. _____ E
            L

    Measure = <u>1 6/10 = 1 3/5 cm or 1.6 cm</u>

Measure each angle in degrees, using a protractor, and record each measure.

5.

    Measure = <u>139°</u>

6.

    Measure = <u>22°</u>

7.

    Measure = <u>44°</u>

8. Define *congruent*: <u>Answers will vary, e.g., "Having the same size and shape."</u>

Determining Congruence

Part A

Measure each line segment in each pair, using a centimeter ruler, and record each measurement in the table. Then, determine whether the two line segments in each pair are congruent or noncongruent. Record your answers in the table.

D

1.
V   J

A

Q

2.

C _____ P

S _____

X

3.   O _____ H

U

74

Name: _____

Determining Congruence

Part B

Measure each angle in each pair, using a protractor, and record each measurement in the table. Then, determine whether the two angles in each pair are congruent or noncongruent. Record your answers in the table.

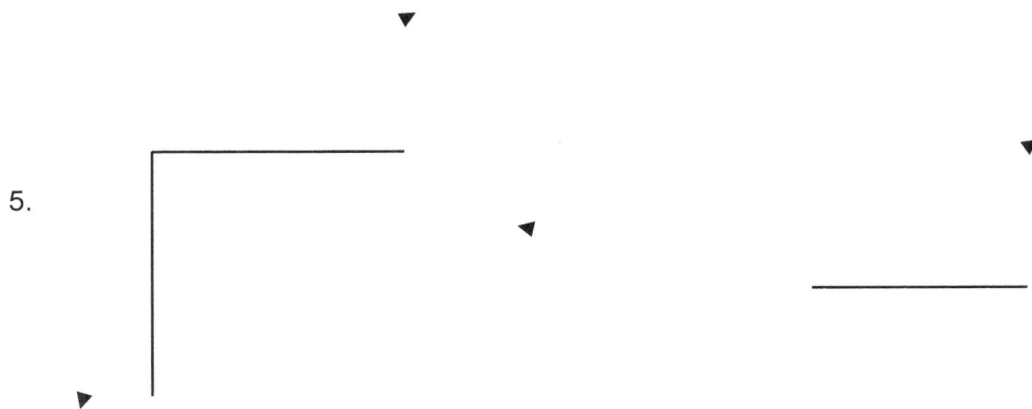

4.

5.

6.

Determining Congruence

Part C

Trace one of the polygons in each pair, using patty paper or tracing paper and a marker. Then, place the tracing over the other polygon in the pair to determine whether the two polygons are congruent or noncongruent. Record your answers in the table.

7.

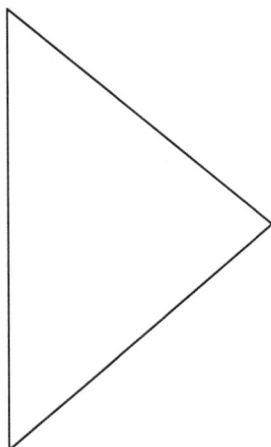 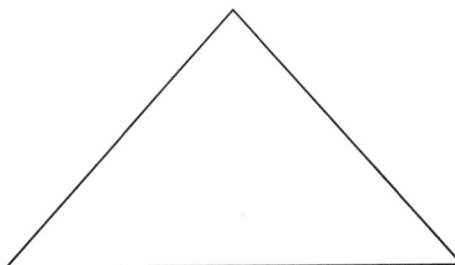

8.

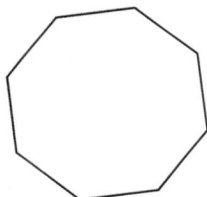 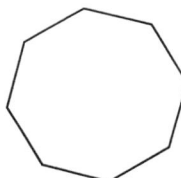

9.

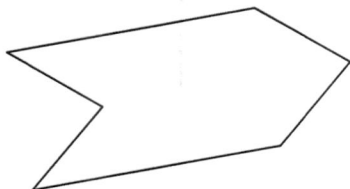 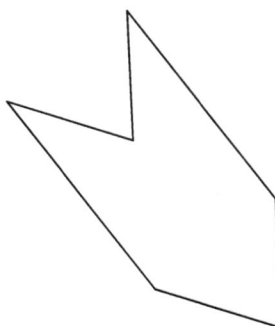

Determining Congruence Tables

Part A

Number	Measure Line Segment 1	Measure Line Segment 2	Congruent or Noncongruent
1			
2			
3			

Part B

Number	Measure Angle	Measure Angle 2	Congruent or Noncongruent
4			
5			
6			

Part C

Number	Congruent or Noncongruent
7	
8	
9	

Name: <u>ANSWER KEY</u>

Determining Congruence Tables

Part A

Number	Measure Line Segment 1	Measure Line Segment 2	Congruent or Noncongruent
1	<u>6 7/10 cm</u> <u>or</u> <u>6.7 cm</u>	<u>6 7/10 cm</u> <u>or</u> <u>6.7 cm</u>	<u>Congruent</u>
2	<u>5 4/10 = 5 2/5 cm</u> <u>or</u> <u>5.4 cm</u>	<u>5 7/10 cm</u> <u>or</u> <u>5.7 cm</u>	<u>Noncongruent</u>
3	<u>4 4/10 = 4 2/5 cm</u> <u>or</u> <u>4.4 cm</u>	<u>4 1/10 cm</u> <u>or</u> <u>4.1</u>	<u>Noncongruent</u>

Part B

Number	Measure Angle	Measure Angle 2	Congruent or Noncongruent
4	<u>45°</u>	<u>74°</u>	<u>Noncongruent</u>
5	<u>134°</u>	<u>134°</u>	<u>Congruent</u>
6	<u>33°</u>	<u>27°</u>	<u>Noncongruent</u>

Part C

Number	Congruent or Noncongruent
7	<u>Congruent</u>
8	<u>Noncongruent</u>
9	<u>Congruent</u>

Reflecting

1. To the right is a practice SOL question. Circle your answer.

   Which figures appear to be congruent?

2. Explain why you chose that answer.

   _____
   _____
   _____
   _____

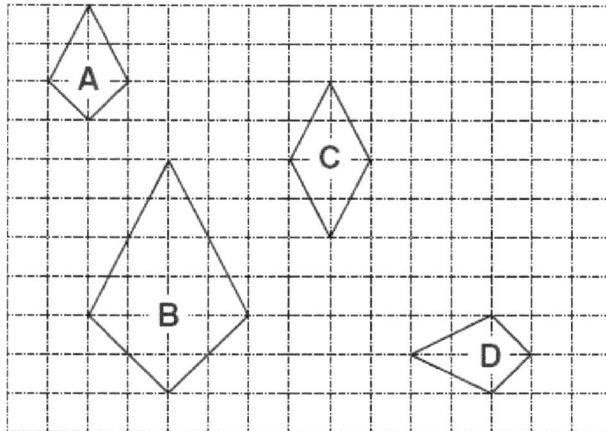

A   A and B
B   B and D
C   C and D
D   D and A

3. Draw two congruent line segments.

4. Draw two congruent angles.

79

Scratch Paper

Reflecting

1. To the right is a practice SOL question. Circle your answer.

    D

2. Explain why you chose that answer.

    Answers will vary, e.g. "A and D are the same size." "A and D are made up of the same number of grid boxes."

Which figures appear to be congruent?

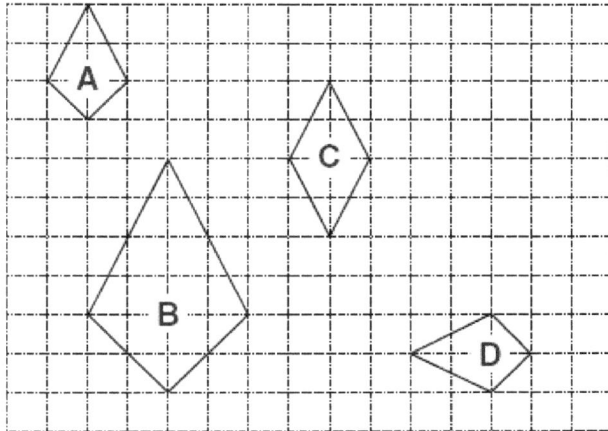

A A and B
B B and D
C C and D
(D) D and A

3. Draw two congruent line segments.

81

Name: _____

Quadrilateral Concept Card

Polygons

These figures are polygons:

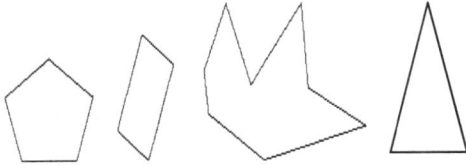

These figures are <u>not</u> polygons:

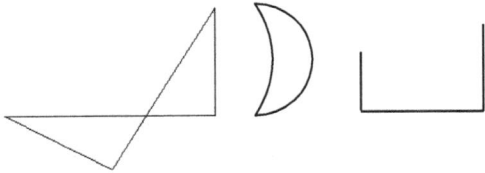

Which of these figures are polygons? (circle)

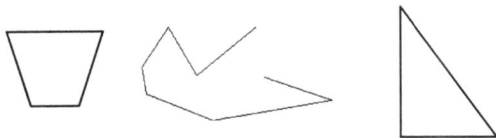

Draw your own example of a polygon.

Draw your own example of a non-polygon.

What is a polygon?

<u>A polygon is</u> _____

_____ .

Quadrilaterals

These figures are quadrilaterals:

These figures are <u>not</u> quadrilaterals:

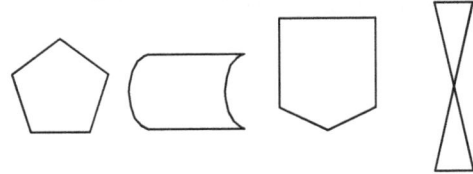

Which of these figures are quadrilaterals? (circle)

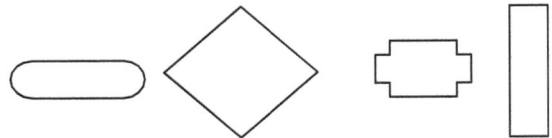

Draw your own example of a quadrilateral.

Draw your own example of a non-quadrilateral.

What is a quadrilateral?

<u>A quadrilateral is</u> _____

_____ .

Scratch Paper

Name: ANSWER KEY

Quadrilateral Concept Card

Polygons

These figures are polygons:

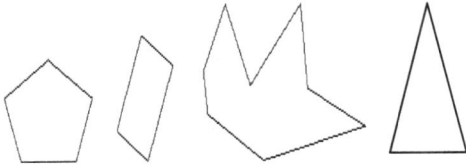

These figures are <u>not</u> polygons:

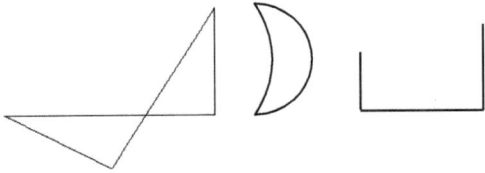

Which of these figures are polygons? (circle)

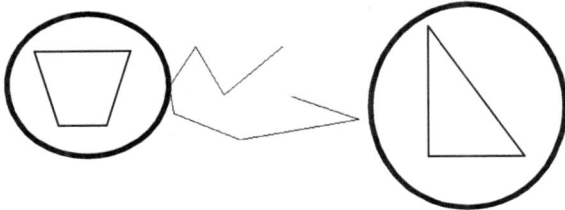

Draw your own example of a polygon.

<u>Drawing will vary.</u>

Draw your own example of a non-polygon.

<u>Drawing will vary. Sample answer:</u>

What is a polygon? <u>Sample answer:</u>

<u>A polygon is a simple, closed, plane figure formed by three or more straight lines.</u>

Quadrilaterals

These figures are quadrilaterals:

These figures are <u>not</u> quadrilaterals:

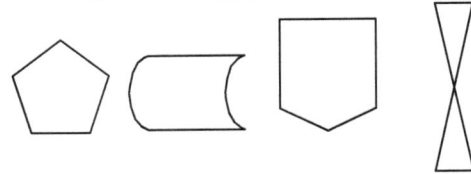

Which of these figures are quadrilaterals? (circle)

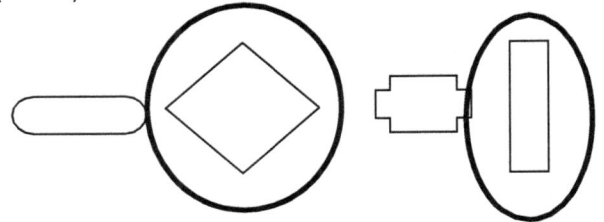

Draw your own example of a quadrilateral.

<u>Drawing will vary.</u>

Draw your own example of a non-quadrilateral.

<u>Drawing will vary. Sample answer:</u>

What is a quadrilateral? <u>Sample answer:</u>

<u>A quadrilateral is a four-sided polygon.</u>

Types of Quadrilaterals

A **quadrilateral** is a closed plane figure with **four sides**.

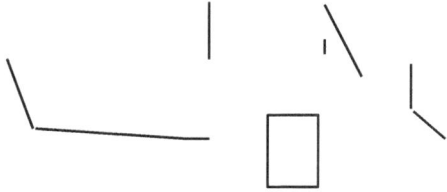

A **parallelogram** is a **quadrilateral** whose **opposite sides are parallel**.

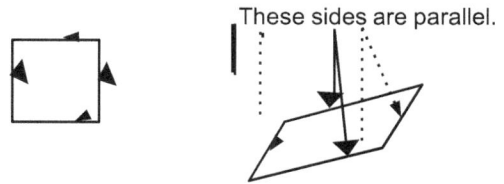

These sides are parallel.

A **rectangle** is a **parallogram** with **four right angles**.

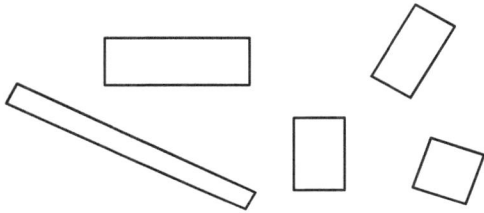

A **rhombus** is a **parallelogram** with **four congruent sides**.

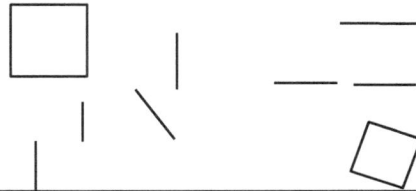

A **square** is a **rectangle** with **four congruent sides** or a **rhombus** with **four right angles**.

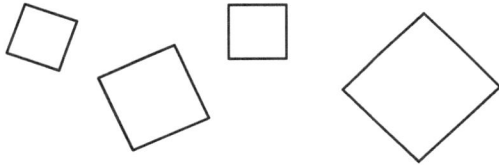

A **trapezoid** is a quadrilateral with **exactly one pair of parallel sides**.

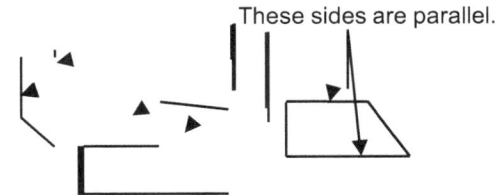

These sides are parallel.

Name: _____

Warm-up

For each question, draw a geometric shape that has all of the properties listed. Write the name of the shape on the line provided.

1. Opposite sides are congruent.
   Opposite sides are parallel.
   Opposite angles are congruent.

   _____

2. All sides are congruent.
   Opposite sides are parallel.
   Opposite angles are equal.
   No angles are equal to 90°.

   _____

3. All angles are right angles.
   Opposite sides are congruent.
   Opposite sides are parallel.

   _____

Scratch Paper

Name: ANSWER KEY

Warm-up

For each question, draw a geometric shape that has all of the properties listed. Write the name of the shape on the line provided.

1. Opposite sides are congruent.
    Opposite sides are parallel.
    Opposite angles are congruent.

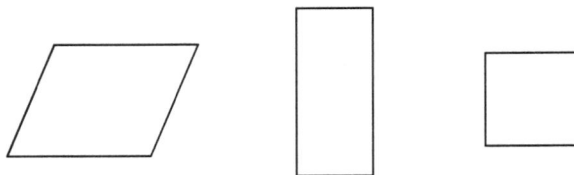

    Sample answers: parallelogram, rectangle, square

2. All sides are congruent.
    Opposite sides are parallel.
    Opposite angles are equal.
    No angles are equal to 90°.

    Sample answers: rhombus, parallelogram

3. All angles are right angles.
    Opposite sides are congruent.
    Opposite sides are parallel.

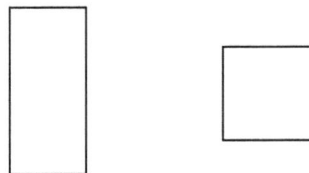

    Sample answers: rectangle, square

Quadrilateral Sorting Pieces

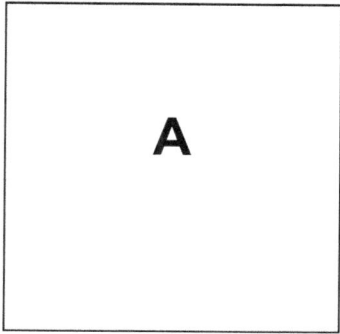

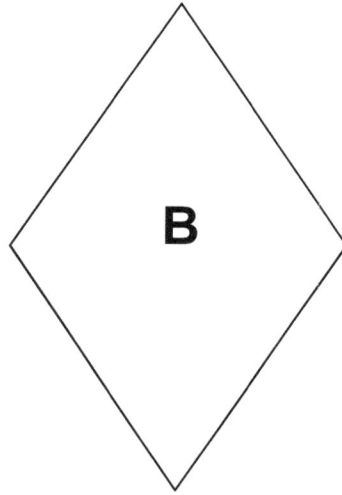

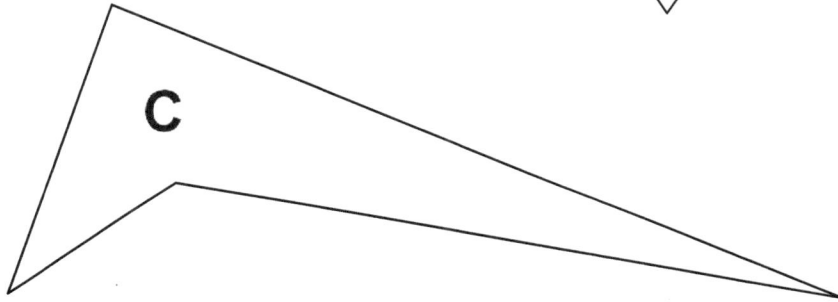

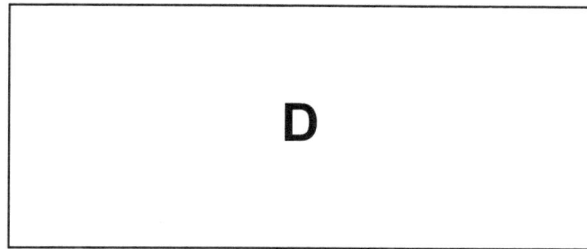

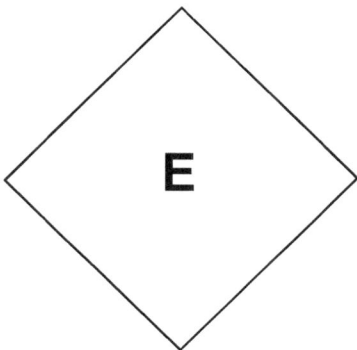

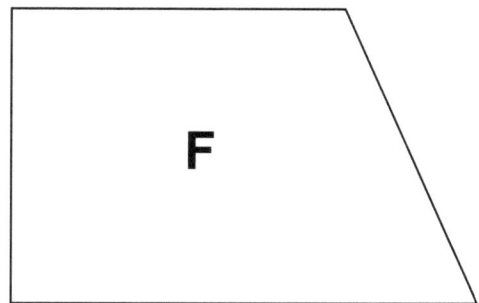

A

B

C

D

E

F

Quadrilateral Sorting Pieces

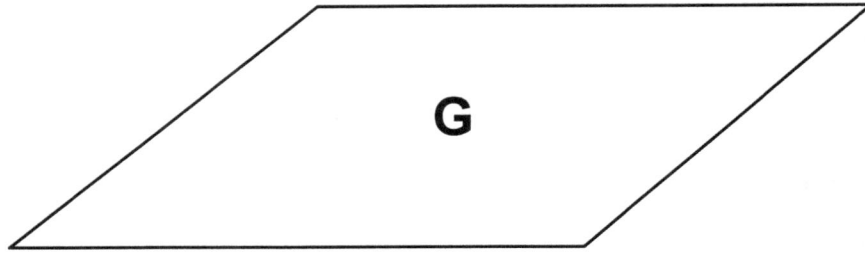

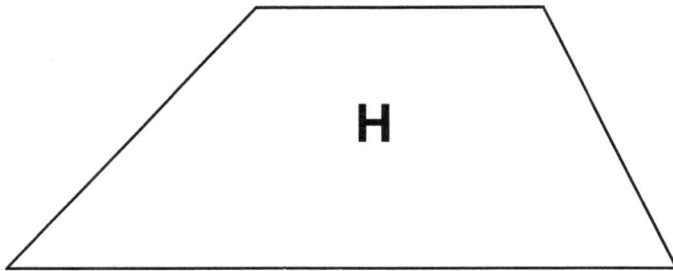

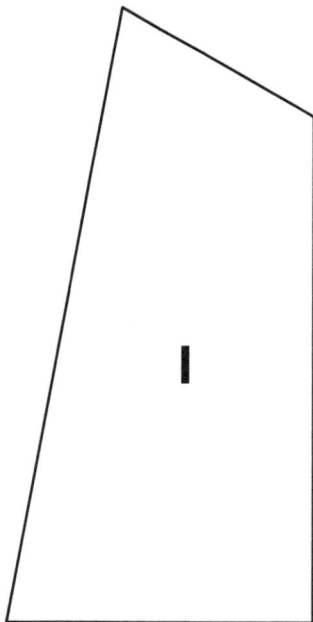

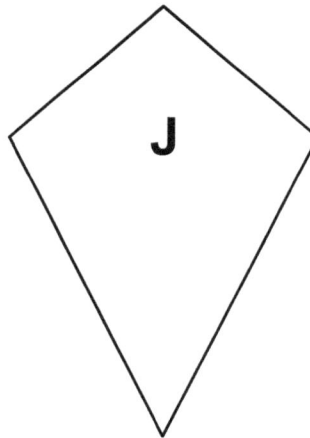

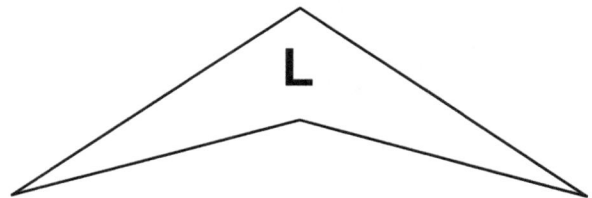

G

H

K

I

J

L

Name: _____

Quadrilateral Activity

Spread out your quadrilateral pieces with the letters facing up so you can see them. Find all of the quadrilaterals that have four right angles, and list them by letter alphabetically on the Question #1 answer line. Then, consider all of the quadrilaterals again. Find all of the quadrilaterals that have exactly one pair of parallel sides, and list them by letter alphabetically in Question #2 answer line. Continue in this manner until you complete all the questions.

1. Has four right angles: _____

2. Has exactly one pair of parallel sides: _____

3. Has two pairs of congruent opposite sides: _____

4. Has four congruent sides: _____

5. Has two pairs of parallel opposite sides: _____

6. Has two pairs of congruent adjacent sides, but not all sides are congruent: _____

7. Has congruent opposite angles: _____

8. Does not have four sides: _____

9. Has four congruent angles: _____

10. Which lists are the same?_____What name can be used to describe quadrilaterals with these properties? _____

Scratch Paper

Quadrilateral Activity

Spread out your quadrilateral pieces with the letters facing up so you can see them. Find all of the quadrilaterals that have four right angles, and list them by letter alphabetically on the Question #1 answer line. Then, consider all of the quadrilaterals again. Find all of the quadrilaterals that have exactly one pair of parallel sides, and list them by letter alphabetically in Question #2 answer line. Continue in this manner until you complete all the questions.

1. Has four right angles: A, D, E, K

2. Has exactly one pair of parallel sides: F, H

3. Has two pairs of congruent opposite sides: A, B, D, E, G, K

4. Has four congruent sides: A, B, E

5. Has two pairs of parallel opposite sides: G, K, A, B, D, E

6. Has two pairs of congruent adjacent sides, but not all sides are congruent: L, J

7. Has congruent opposite angles: A, B, D, E, G, K

8. Does not have four sides: no examples

9. Has four congruent angles: A, D, E, K

10. Which lists are the same? #3, 5, and 8   What name can be used to describe quadrilaterals with these properties? parallelograms

Quadrilateral Family Tree

Fill out the family tree by writing the names *quadrilateral, rectangle, square, rhombus, parallelogram,* and *trapezoid* into the appropriate blocks on the diagram.

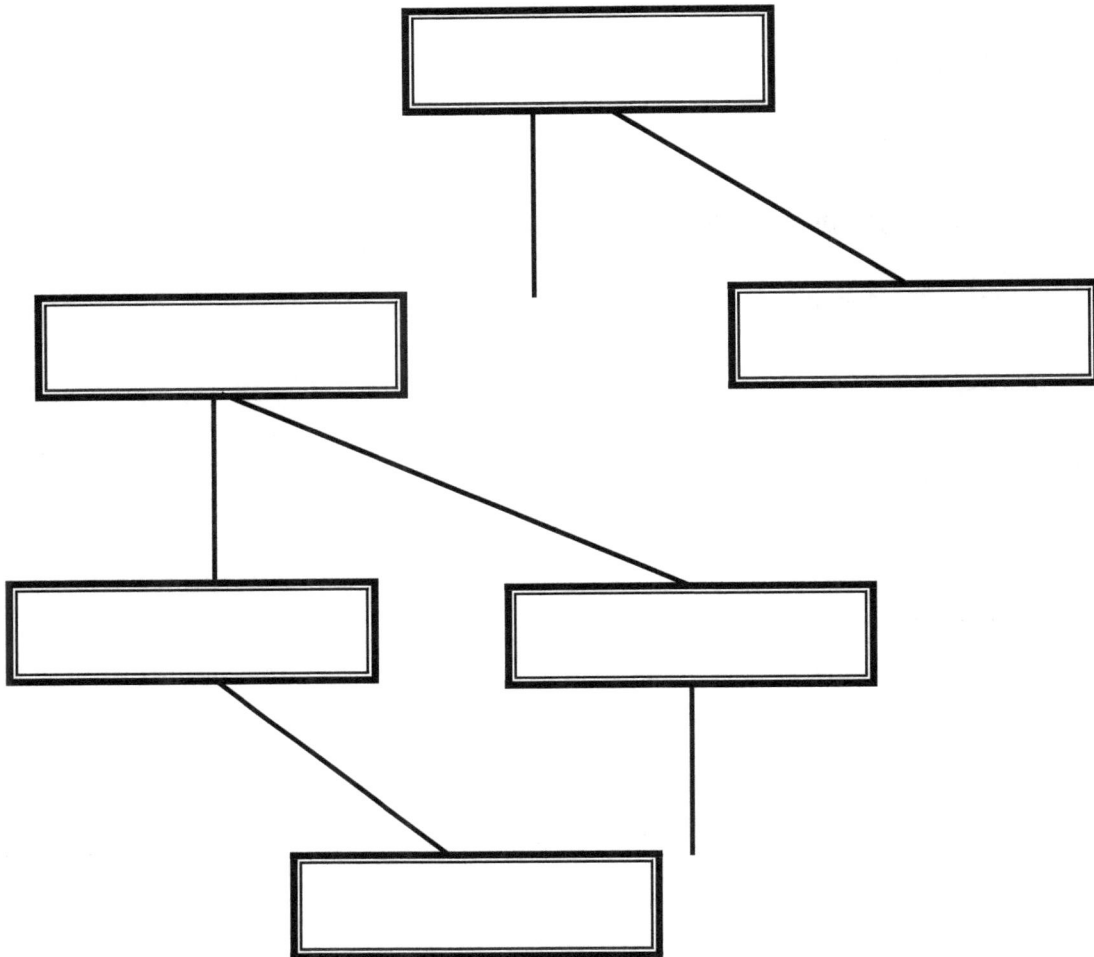

Scratch Paper

Name: <u>ANSWER KEY</u>

Quadrilateral Family Tree

Fill out the family tree by writing the names *quadrilateral, rectangle, square, rhombus, parallelogram,* and *trapezoid* into the appropriate blocks on the diagram.

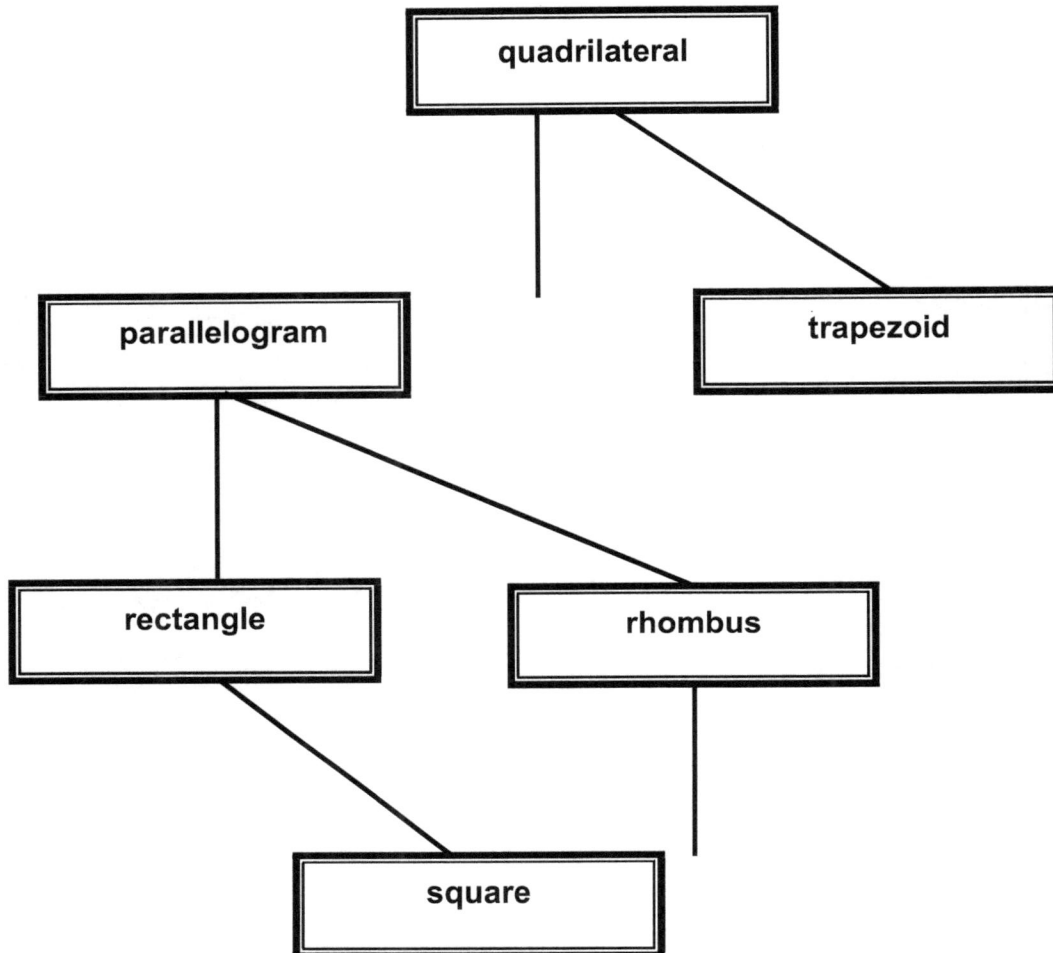

```
                    ┌─────────────────┐
                    │  quadrilateral  │
                    └─────────────────┘
                       /          \
   ┌─────────────────┐          ┌─────────────┐
   │  parallelogram  │          │  trapezoid  │
   └─────────────────┘          └─────────────┘
        /        \
┌─────────────┐  ┌─────────────┐
│  rectangle  │  │   rhombus   │
└─────────────┘  └─────────────┘
        \          /
      ┌─────────────┐
      │   square    │
      └─────────────┘
```

Name: _____

Quadrilateral Puzzle

Match the numbered vocabulary word with the block in the puzzle that shows an example of that word. Place the letter or symbol from the puzzle block onto the numbered line below to solve the puzzle. Each block may be used only once even though the shapes in some blocks have more than one name.

1. square
2. rectangle
3. right angle
4. parallel lines
5. rhombus
6. trapezoid
7. parallelogram
8. quadrilateral
9. polygon

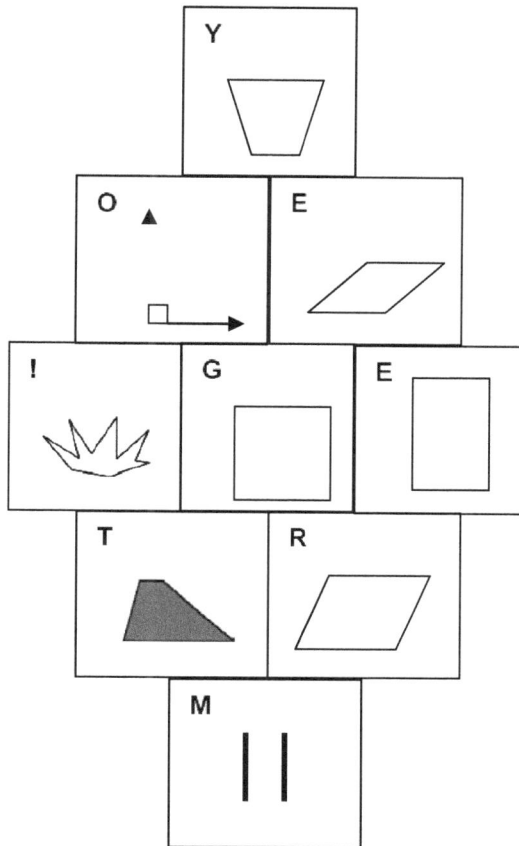

Y

O ▲

E

!

G

E

T

R

M ||

1 2 3 4 5 6 7 8 9

97

Scratch Paper

Quadrilateral Puzzle

Match the numbered vocabulary word with the block in the puzzle that shows an example of that word. Place the letter or symbol from the puzzle block onto the numbered line below to solve the puzzle. Each block may be used only once even though the shapes in some blocks have more than one name.

1. square

2. rectangle

3. right angle

4. parallel lines

5. rhombus

6. trapezoid

7. parallelogram

8. quadrilateral

9. polygon

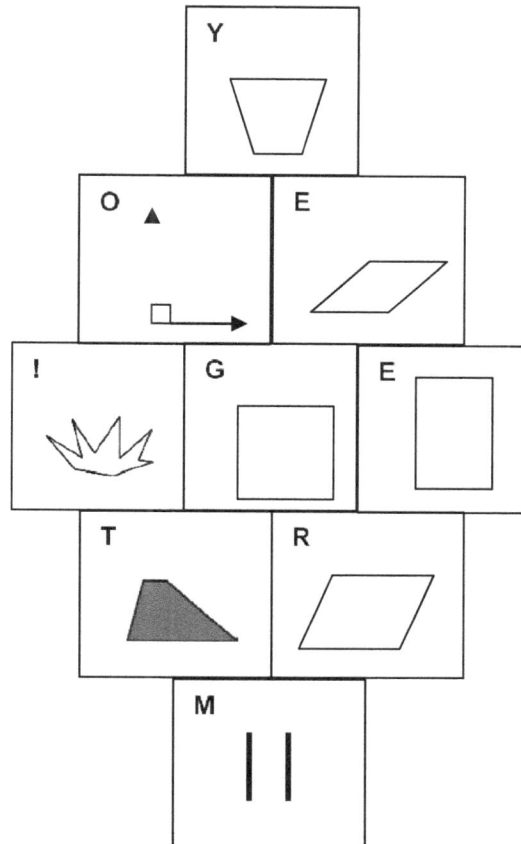

<u>G</u> <u>E</u> <u>O</u> <u>M</u> <u>E</u> <u>T</u> <u>R</u> <u>Y</u> <u>!</u>
1   2   3   4   5   6   7   8   9

99

Name: _____

Alphabet Similarity

Determine whether the pairs of upper and lower case letters are similar or not.

Aa   Bb   Cc   Dd   Ee

Ff   Gg   Hh   Ii   Jj

Kk   Ll   Mm   Nn   Oo

Pp   Qq   Rr   Ss   Tt

Uu   Vv   Ww   Xx   Yy

Zz

Scratch Paper

Alphabet Similarity

Determine whether the pairs of upper and lower case letters are similar or not.

Aa   **Bb**   **Cc**   **Dd**   **Ee**

Similar (under Cc)

Ff   **Gg**   **Hh**   **Ii**   **Jj**

Kk   **Ll**   **Mm**   **Nn**   **Oo**

Similar (under Kk)   Similar (under Oo)

Pp   **Qq**   **Rr**   **Ss**   **Tt**

Similar (under Pp)   Similar (under Ss)

Uu   **Vv**   **Ww**   **Xx**   **Yy**

Similar   Similar   Similar   Similar

Zz

Similar

Name: _____

Reflecting on Similar Figures

Four triangles are shown on the grid below.

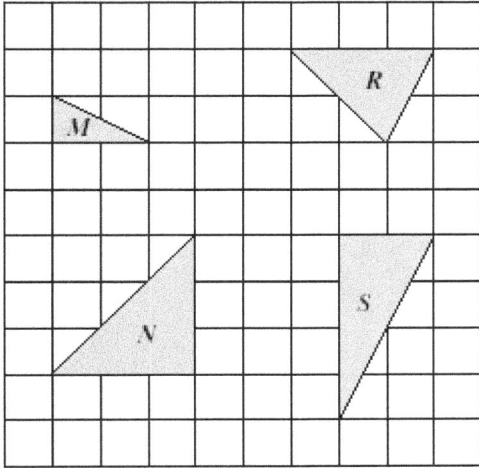

Which two triangles appear to be similar?

F   *M* and *S*
G   *M* and *N*
H   *N* and *S*
J   *R* and *N*

In the first column below, draw a simple figure. In the second column, draw a figure that is *similar* to the first figure.

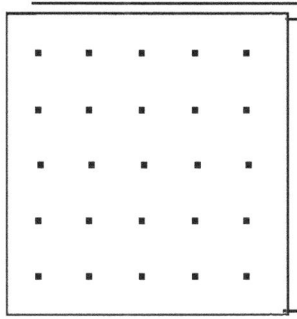 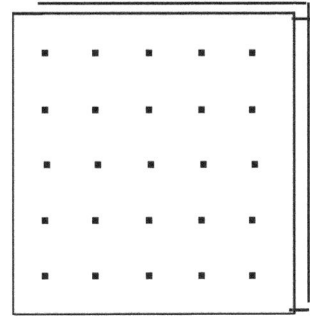

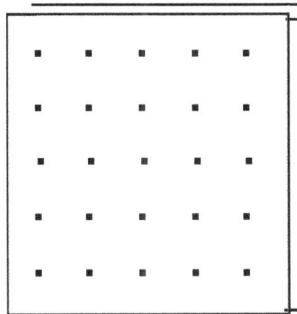 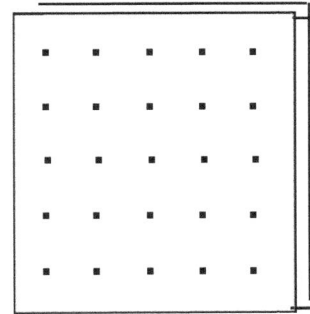

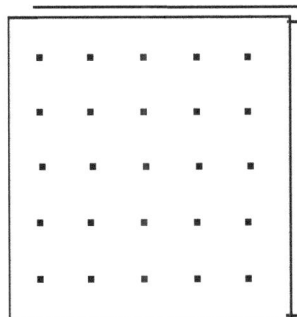 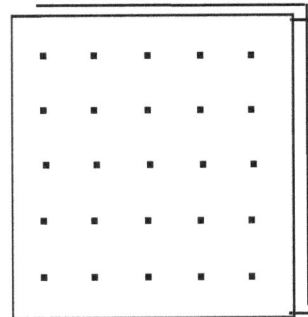

Scratch Paper

Reflecting on Similar Figures

Four triangles are shown on the grid below.

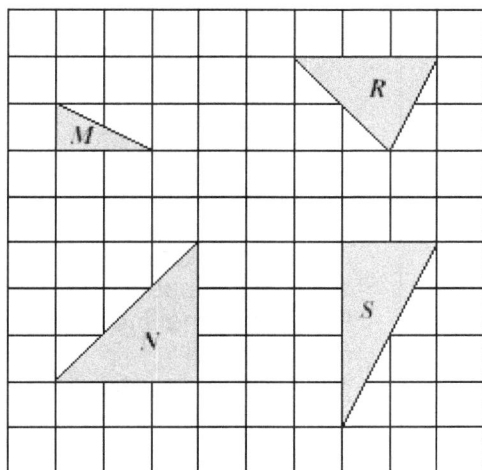

Which two triangles appear to be similar?

F M and S

G M and N

H N and S

J R and N

In the first column below, draw a simple figure. In the second column, draw a figure that is *similar* to the first figure.

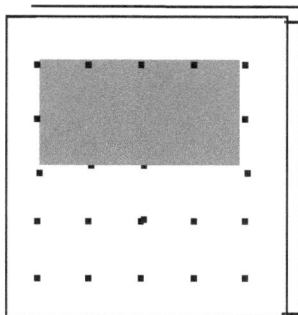 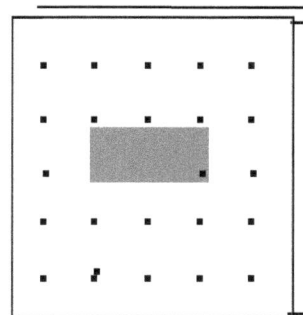

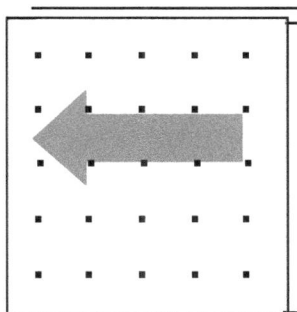 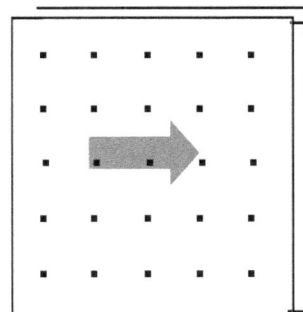

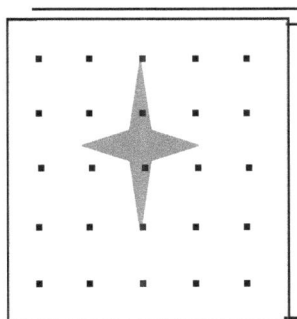 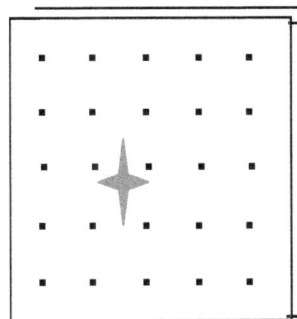

Transformation Definitions Chart

Translation (slide)

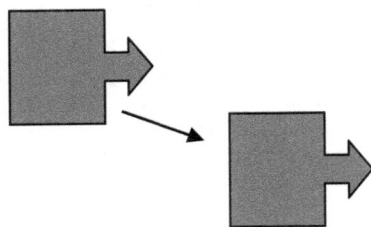

A transformation in which an image is formed by moving every point on a figure the same distance in the same direction.

Rotation (turn)

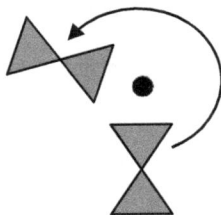

A transformation in which an image is formed by turning its pre-image around a fixed point.

Reflection (flip)

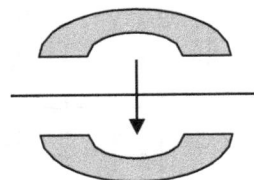

A transformation in which a figure is flipped over a line, called the line of reflection. All corresponding points in the image and pre-image are equidistant from the line of reflection.

Translation, Reflection, Rotation

Trace each original figure. Use the tracing to perform a translation, reflection, and rotation, and draw the result of each in the appropriate box.

Original Figure	Translation	Reflection	Rotation

Scratch Paper

Translation, Reflection, Rotation

Trace each original figure. Use the tracing to perform a translation, reflection, and rotation, and draw the result of each in the appropriate box.
<u>Sample answers:</u>

Original Figure	Translation	Reflection	Rotation

Name: _____

Move Those Shapes!

Draw all three transformations discussed in this lesson: **translation, reflection,** and **rotation.** Draw a figure in the top left portion of the grid paper. The figure should be different from all the others used in this lesson. Perform the first transformation on the original figure, and label the new figure "Figure B." Perform the second transformation on Figure B, and label the new figure "Figure C." Perform the final transformation on Figure C, and label the new figure "Figure D." Draw arrows to show the direction of each transformation.

Transformations: Fig. B. _____   Fig. C. _____   Fig. D. _____

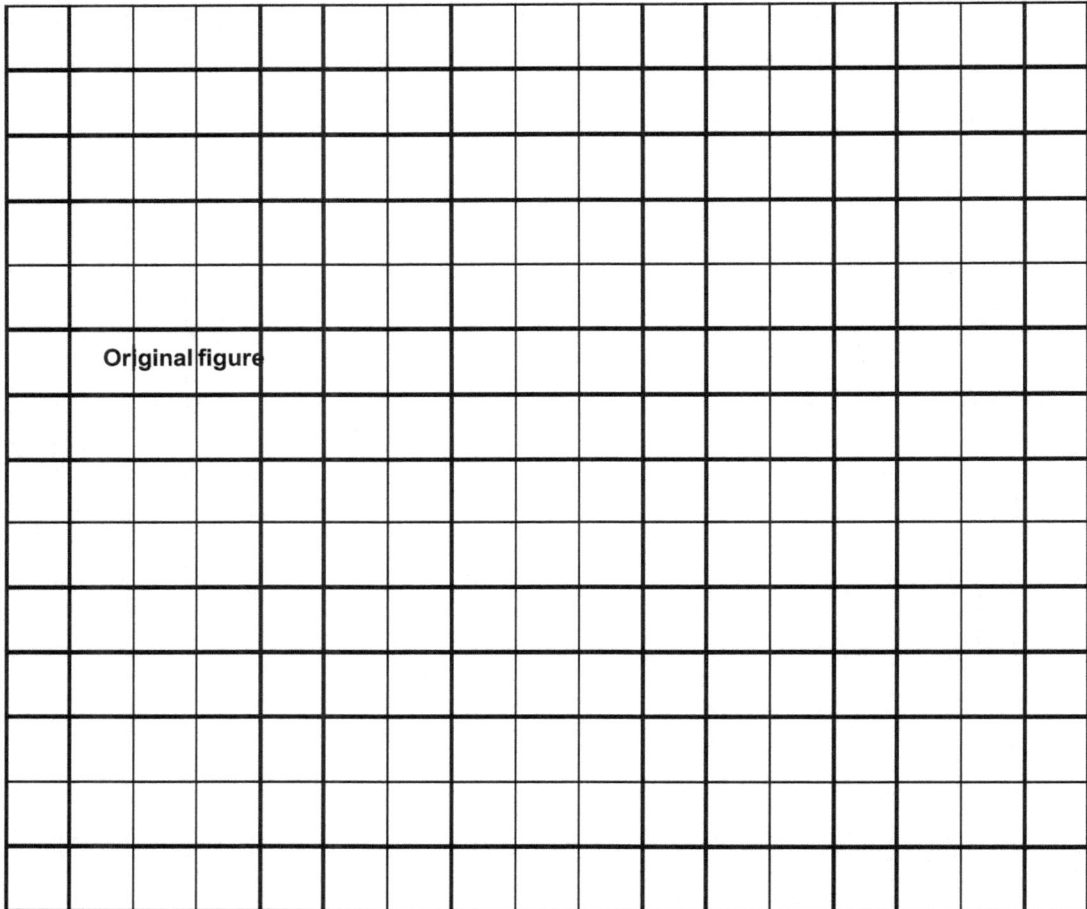

Original figure

Scratch Paper

Move Those Shapes!

Draw all three transformations discussed in this lesson: **translation, reflection,** and **rotation.** Draw a figure in the top left portion of the grid paper. The figure should be different from all the others used in this lesson. Perform the first transformation on the original figure, and label the new figure "Figure B." Perform the second transformation on Figure B, and label the new figure "Figure C." Perform the final transformation on Figure C, and label the new figure "Figure D." Draw arrows to show the direction of each transformation.

Transformations: Fig. B. reflection   Fig. C. rotation   Fig. D. translation

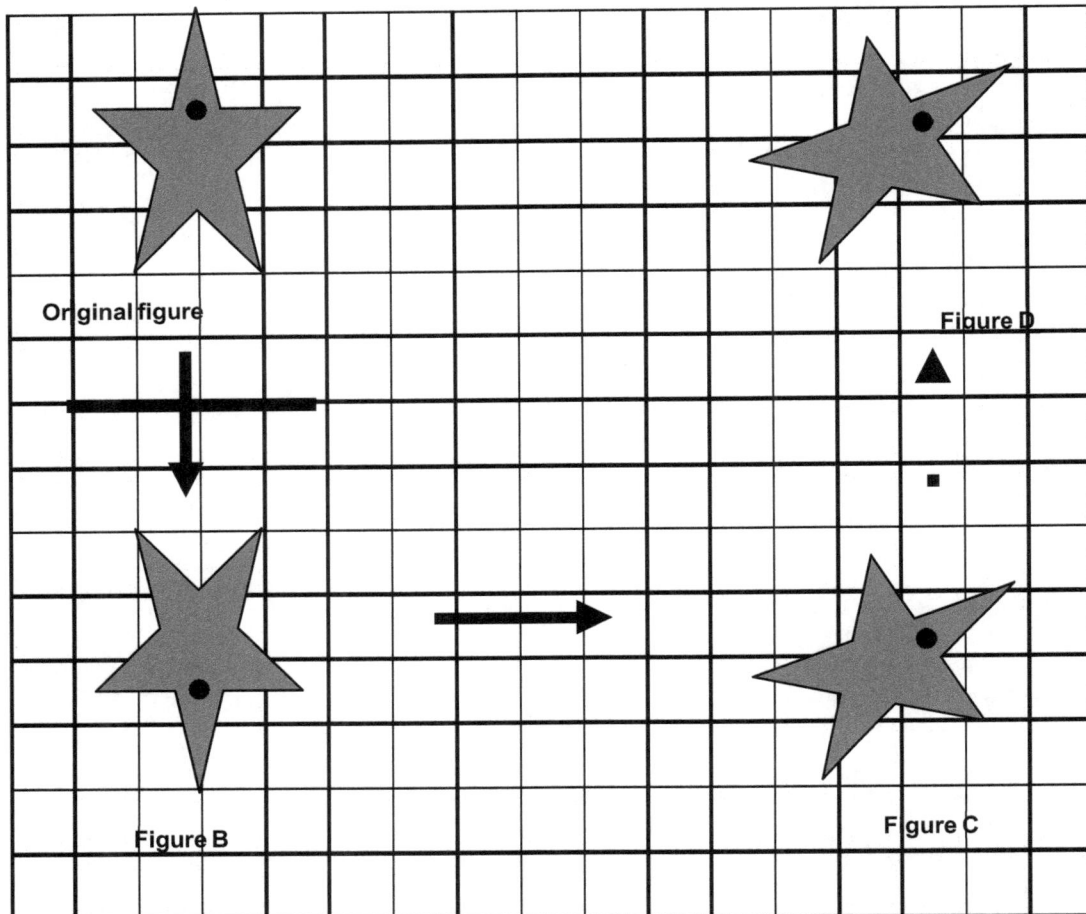

112

Name: _____

Reflecting on Transformations

For each change in position of the figures below, determine the type of transformation it is, and write the name of the transformation on the line provided.

1. The example below is an illustration of a_____.

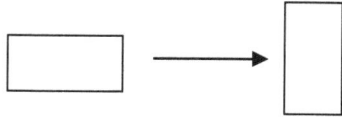

2. The example below is an illustration of a_____.

3. The example below is an illustration of a_____.

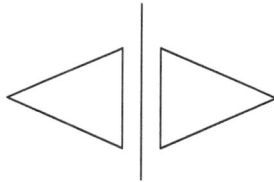

4. The example below is an illustration of a_____.

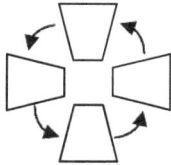

5. Write directions about how to go from your bedroom to your kitchen, using the verbs *translate, reflect,* and *rotate.*

113

Scratch Paper

Reflecting on Transformations

For each change in position of the figures below, determine the type of transformation it is, and write the name of the transformation on the line provided.

1. The example below is an illustration of a <u>rotation</u>.

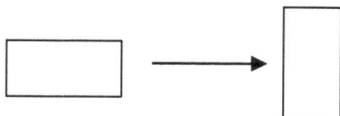

2. The example below is an illustration of a <u>translation</u>.

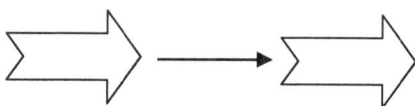

3. The example below is an illustration of a <u>reflection</u>.

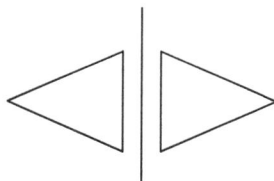

4. The example below is an illustration of a <u>rotation</u>.

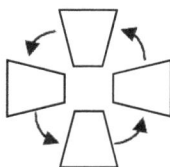

5. Write directions of how to go from your bedroom to your kitchen, using translations, reflections, and rotations.

    <u>Sample Answer:</u>

    <u>Rotate to the right out my door, translate past two doors, rotate to the left, translate down the stairs, and reflect into the kitchen.</u>

Name: _____

Warm-up

1. Label and number the *x*-axis and *y*-axis.

2. Graph the following points, and connect the dots as you go: (−3, −1), (2, −1), (4, 1), (−1, 1)

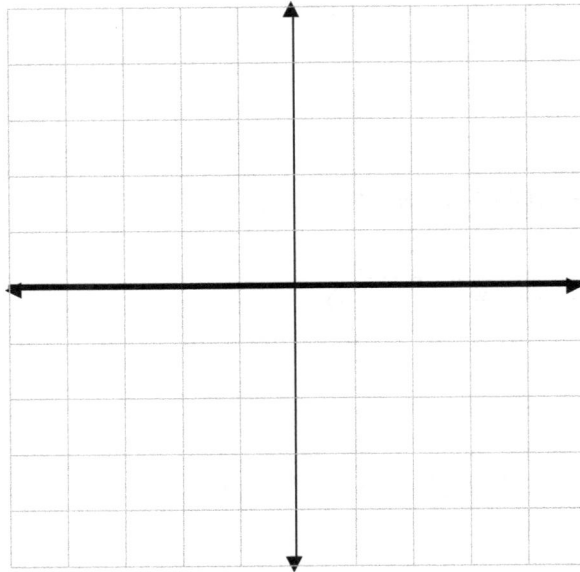

3. What shape did you just draw? _____

Scratch Paper

Name: ANSWER KEY

Warm-up

1. Label and number the *x*-axis and *y*-axis.

2. Graph the following points, and connect the dots as you go: (−3, −1), (2, −1), (4, 1), (−1, 1)

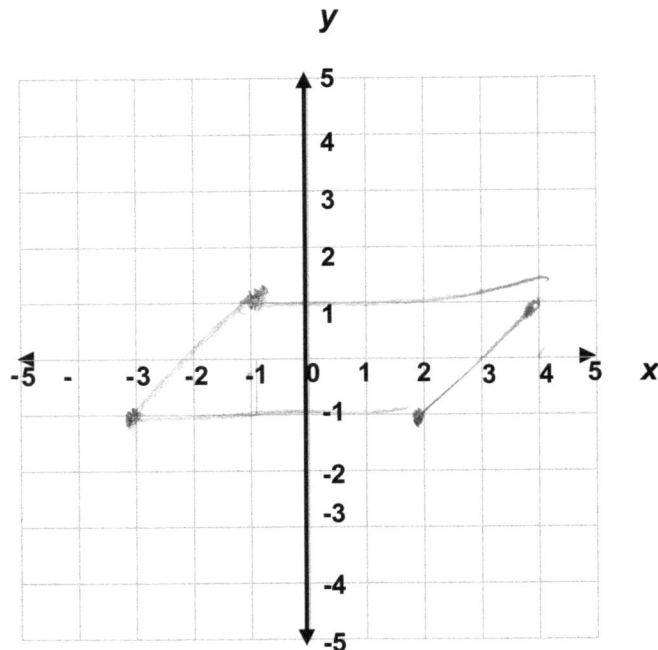

3. What shape did you just draw? <u>parallelogram</u>

Translations

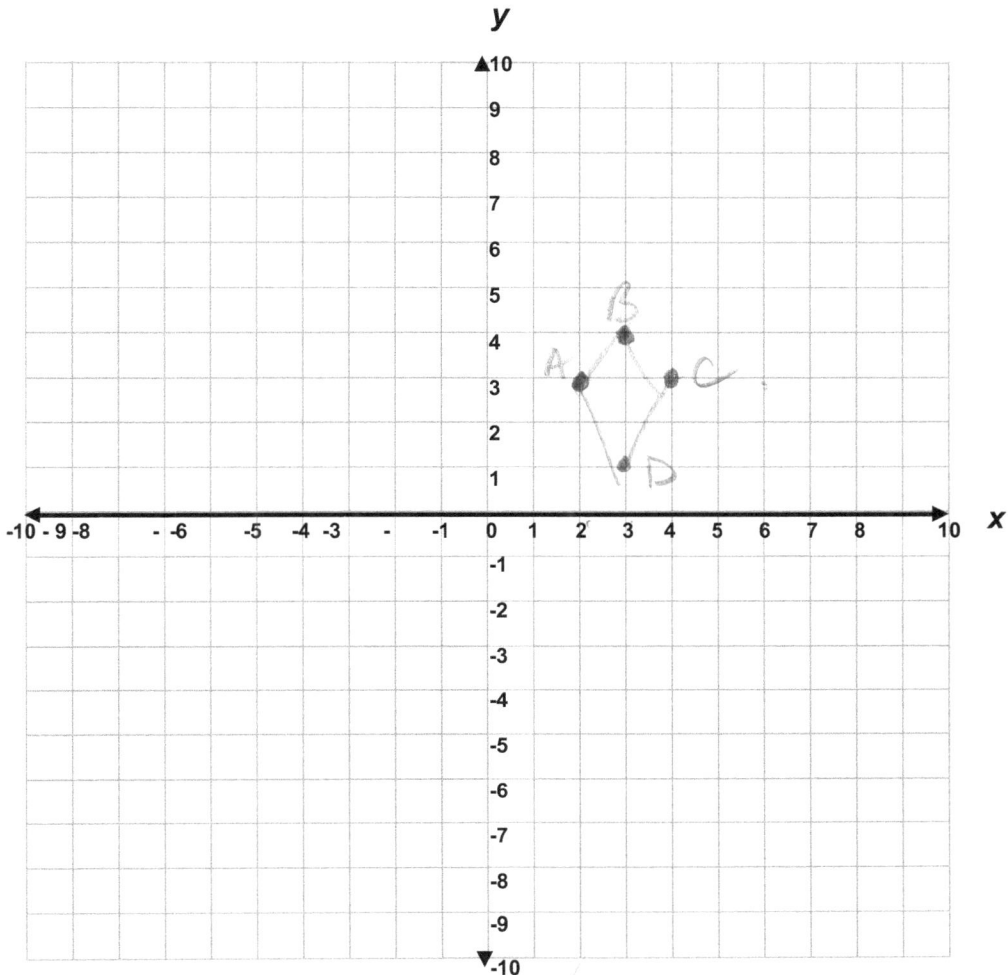

Enter the new coordinates after performing each translation shown below:

IMAGE	ORIGINAL COORDINATES			
	A (2, 3)	B (3, 4)	C (4, 3)	D (3, 1)
Image 1: 4 units right	6, 3	7, 4	8, 3	7, 1
Image 2: 7 units left	-5, 3	-4, 4	-3, 3	-4, 1
Image 3: 5 units up	2,			
Image 4: 8 units down				
Image 5: 5 units right *and* 4 units up	7, 7			
Image 6 (coordinates of choice in the 3rd quadrant):				

119

Scratch Paper

Translations

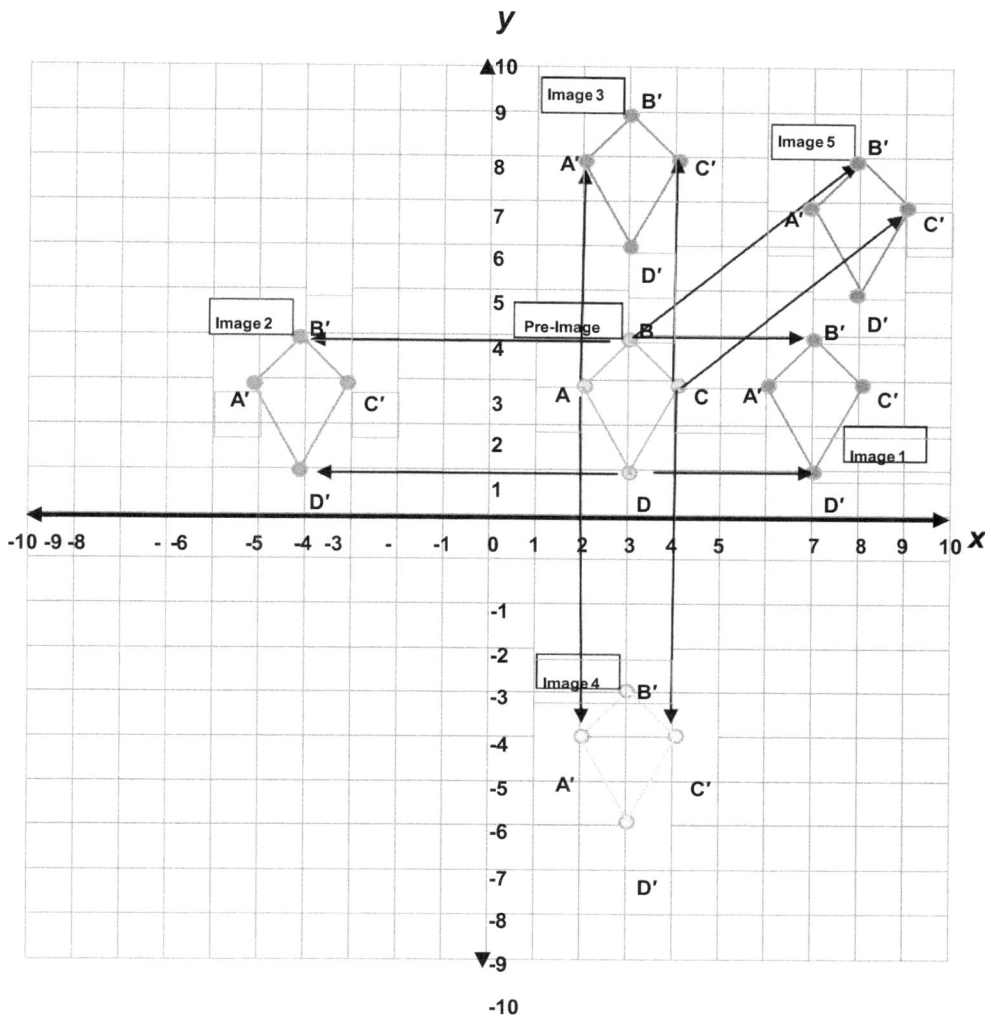

Enter the new coordinates after performing each translation shown below:

IMAGE	ORIGINAL COORDINATES			
	A (2, 3)	B (3, 4)	C (4, 3)	D (3, 1)
Image 1: 4 units right	(6, 3)	(7, 4)	(8, 3)	(7, 1)
Image 2: 7 units left	(−5, 3)	(−4, 4)	(−3, 3)	(−4, 1)
Image 3: 5 units up	(2, 8)	(3, 9)	(4, 8)	(3, 6)
Image 4: 8 units down	(2, −5)	(3, −4)	(4, −5)	(3, −7)
Image 5: 5 units right *and* 4 units up	(7, 7)	(8, 8)	(9, 7)	(8, 5)
_____				

121

Name: _____

Reflecting on Translations

Each figure below shows a transformation. The pre-image is shown with solid lines and the image is shown with dashed lines. Next to each figure, tell whether the transformation shown represents a translation, and then explain why or why not.

1.

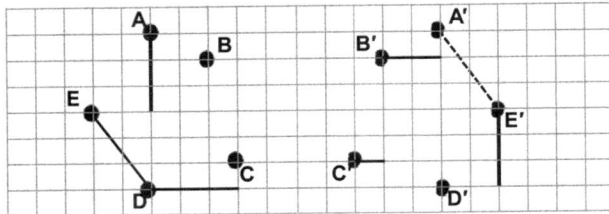

Is this a translation? _____
Explain:

_____

_____

2.

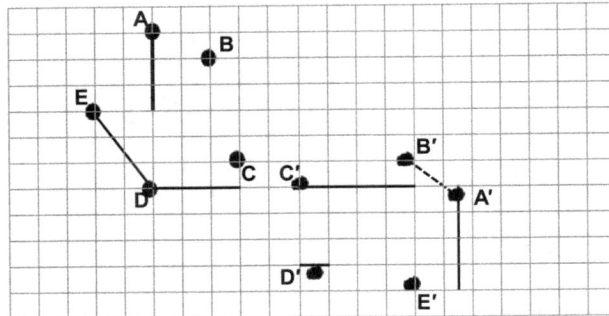

Is this a translation? _____
Explain:

_____

_____

3.

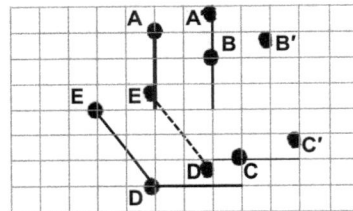

Is this a translation? _____
Explain:

_____

_____

4.

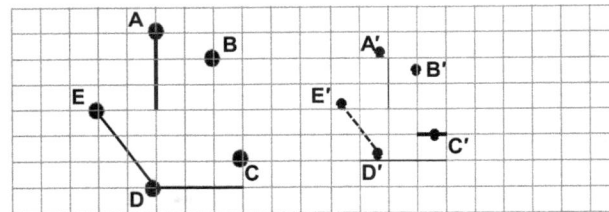

Is this a translation? _____
Explain:

_____

_____

5. When you translate a figure on the coordinate plane, explain how the x-coordinate and y-coordinate of each point is affected.

_____

_____

Scratch Paper

Reflecting on Translations

Each figure below shows a transformation. The pre-image is shown with solid lines and the image is shown with dashed lines. Next to each figure, tell whether the transformation shown represents a translation, and then explain why or why not.

1.

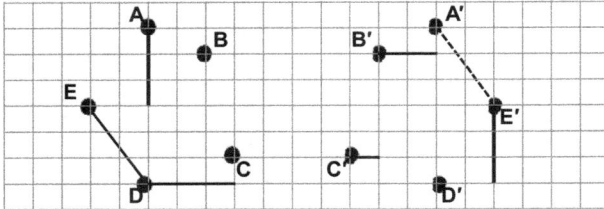

Is this a translation? <u>No</u>
Explain:
<u>The points have moved by different amounts.</u>

2.

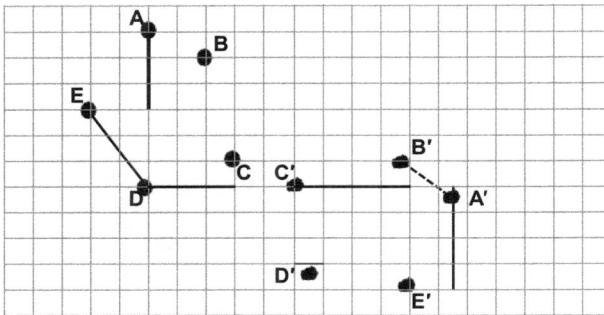

Is this a translation? <u>No</u>
Explain:
<u>The figure has rotated. Also, the points have moved by different amounts.</u>

3.

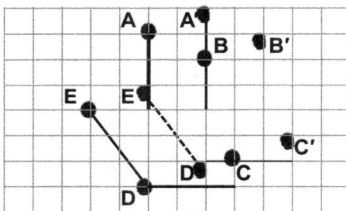

Is this a translation? <u>Yes</u>
Explain:
<u>All the points have moved right 2 units and up 1 unit. The figure has not rotated.</u>

4.

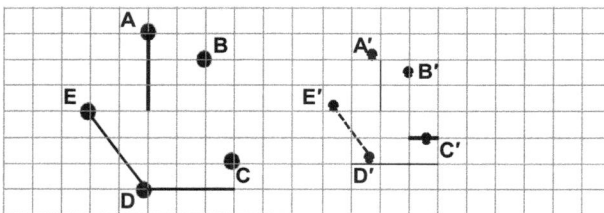

Is this a translation? <u>No</u>
Explain:
<u>The figure has changed in size. In a translation, the figure remains the same size.</u>

5. When you translate a figure on the coordinate plane, explain how the x-coordinate and y-coordinate of each point is affected.
<u>Moving a figure left or right causes the x-coordinate to change. Moving a figure up or down causes the y-coordinate to change. (Actual responses may vary.)</u>

6. What happens to the graph of a figure (in what direction and how far does it move) when...
   - 3 is added to the x-coordinate? <u>It moves 3 units to the right.</u>
   - 10 is added to the y-coordinate? <u>It moves 10 units up.</u>
   - 7 is subtracted from the y-coordinate? <u>It moves 7 units down.</u>
   - 6 is subtracted from the x-coordinate? <u>It moves 6 units to the left.</u>
   - 2 is added to the x-coordinate *and* 3 is subtracted from the y-coordinate? <u>It moves 2 units to the right *and* 3 units down.</u>

Name: _____

Warm-up

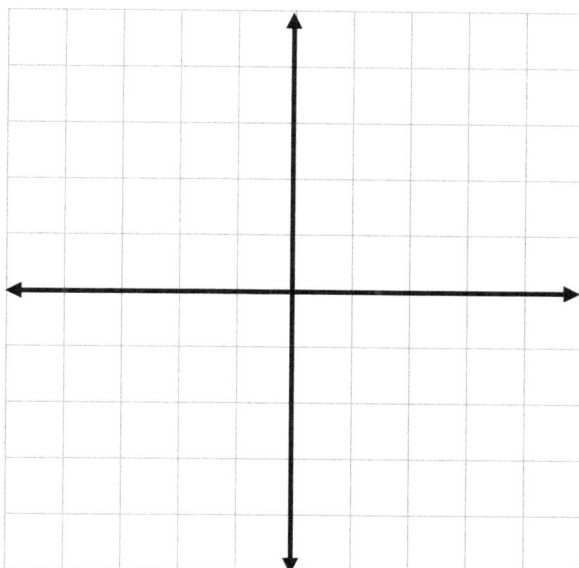

Part I

1. Label and number the x-axis and the y-axis.

2. Graph the following points, and connect them as you go: (2, 0), (−1, −1), (−3, 2) (0, 5), (3, 3).

3. What shape did you just draw? _____

Part II

4. Stand up, face a wall, and hold your arm straight in front of you. While standing in the same spot, turn your body to the left. How many degrees did you just rotate?_____Did you rotate clockwise or counterclockwise? _____

5. Return to your original standing position with your arm straight in front of you. While standing in the same spot, turn your body to right all the way around until you are back in the position where you started. How many degrees did you just rotate?_____Did you rotate clockwise or counterclockwise? _____

6. Write a list of things that ordinarily rotate or turn.

_____

_____

Scratch Paper

Name: **ANSWER KEY**

Warm-up

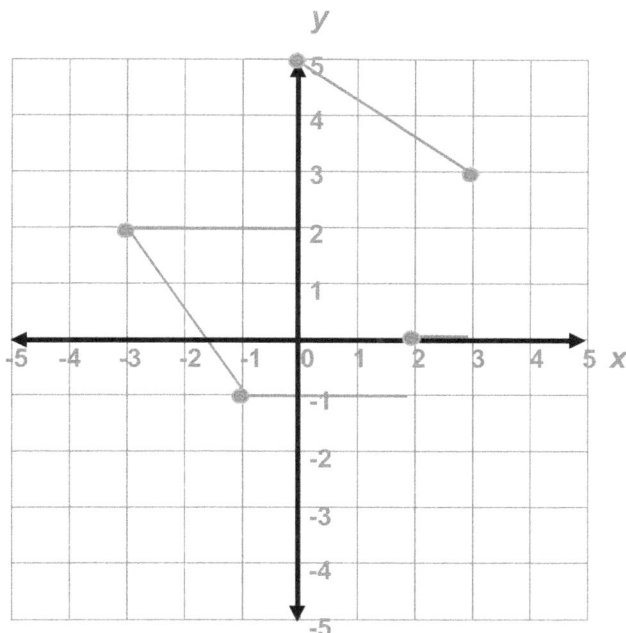

Part I

1. Label and number the x-axis and the y-axis.

2. Graph the following points, and connect them as you go: (2, 0), (−1, −1), (−3, 2) (0, 5),(3, 3).

3. What shape did you just draw?____pentagon_____

Part II

4. Stand up, face a wall, and hold your arm straight in front of you. While standing in the same spot, turn your body to the left. How many degrees did you just rotate?____90°___Did you rotate clockwise or counterclockwise?___counterclockwise___

5. Return to your original standing position with your arm straight in front of you. While standing in the same spot, turn your body to right all the way around until you are back in the position where you started. How many degrees did you just rotate?___360°___Did you rotate clockwise or counterclockwise?___clockwise___

6. Write a list of things that ordinarily rotate or turn.

   Possible responses: tire, merry-go-round, ferris wheel, fan, propeller, a hand on a clock

Name: _____

Rotations

Follow the verbal directions of your teacher to complete this worksheet.

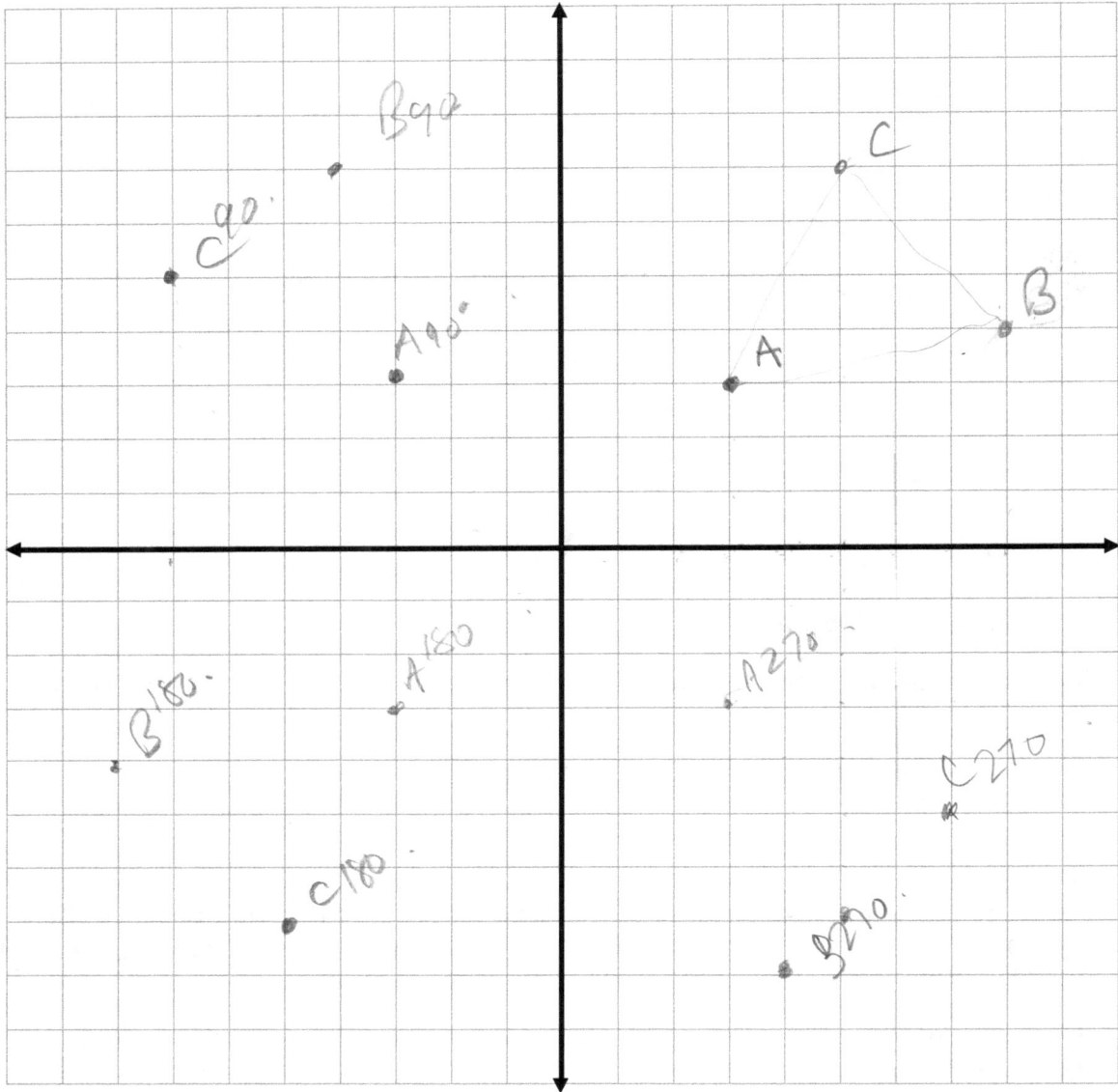

Original Coordinates	90° Rotation	180° Rotation	270° Rotation	360° Rotation
A (3 , 3)	-3, 3	-3, -3	3, -3	3, 3
B (8 , 4)	-4, 8	-8, -4	4, -8	___
C (5 , 7)	-7, 5	-5, -7	7, 5	___

Name: ANSWER KEY

Rotations

Follow the verbal directions of your teacher to complete this worksheet.

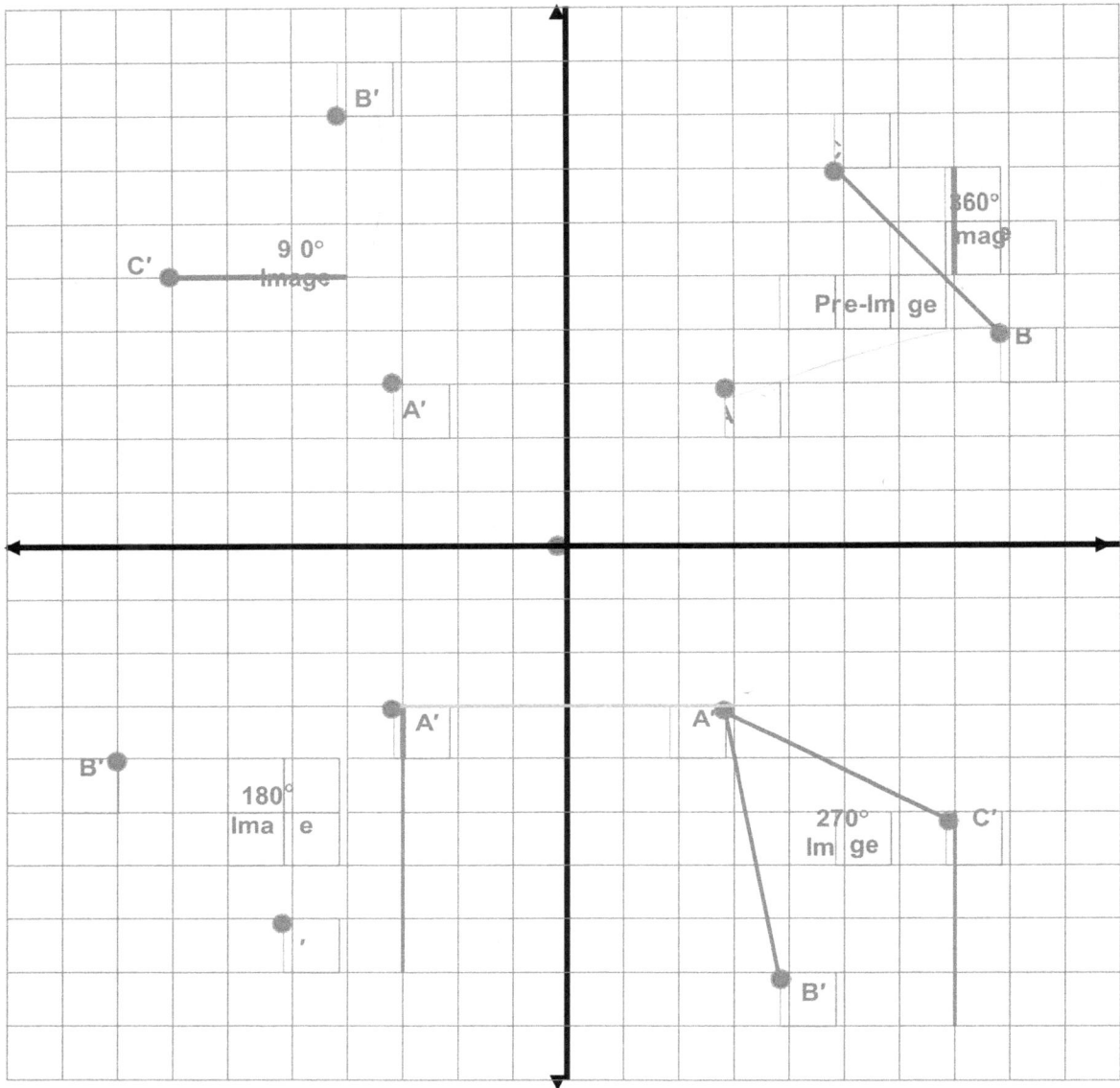

Original Coordinates	90° Rotation	180° Rotation	270° Rotation	360° Rotation
A (3 , 3)	(−3, 3)	(−3, −3)	(3, −3)	(3, 3)
B (8 , 4)	(−4, 8)	(−8, −4)	(4, −8)	(8, 4)
C (5 , 7)	(−7, 5)	(−5, −7)	(7, −5)	(5, 7)

130

Reflecting on Rotations

1. When a figure is rotated on the coordinate plane, does its shape change? _____

2. At right is a sample SOL test question related to rotations. How could you use patty paper to help answer the question?

   _____

   _____

3. Which answer is correct? _____

The diagram below shows a geometric figure on a coordinate plane.

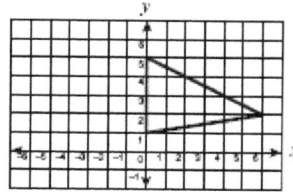

Which of the diagrams below shows a rotation of this geometric figure?

F

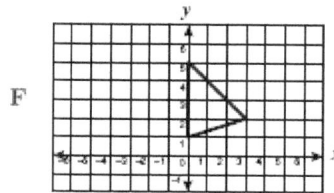

G

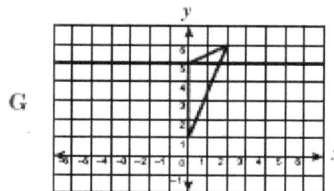

H

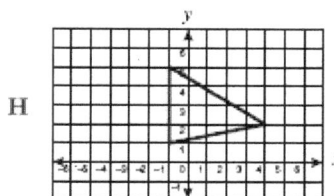

J

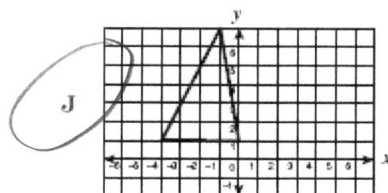

Scratch Paper

Reflecting on Rotations

1. When a figure is rotated on the coordinate plane, does its shape change?_____no_____

The diagram below shows a geometric figure on a coordinate plane.

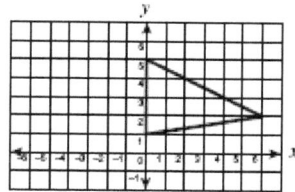

2. At right is a sample SOL test question related to rotations. How could you use patty paper to help answer the question?

   You could trace the original pre-image on the patty paper and then put the patty paper over each answer image, rotating the tracing to see if the answer image is a rotation.

Which of the diagrams below shows a rotation of this geometric figure?

F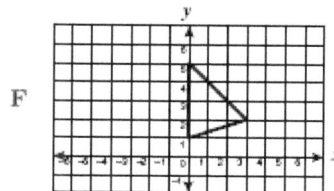

3. Which answer is correct?___J___

G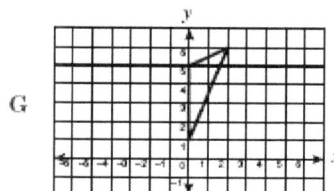

H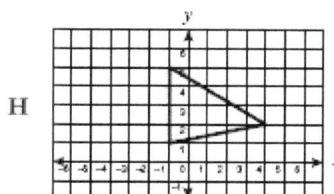

J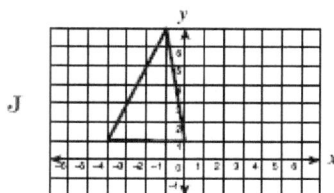

Name: _____

Warm-up

1. Make a list of real-life situations in which you see "scaled up" or "scaled down" versions of objects.

2. Practice multiplying a fraction by a whole number:

    a. $\frac{1}{2} \cdot 6 =$

    b. $\frac{1}{3} \cdot 12 =$

    c. $\frac{3}{5} \cdot 10 =$

    d. $\frac{1}{4} \cdot 0 =$

Name: ANSWER KEY

Warm-up

1. Make a list of real-life situations in which you see "scaled up" or "scaled down" versions of objects.

   Answers will vary. Possible responses: maps, blueprints, model airplanes, architectural models, toys that are scaled down versions of real objects, images taken from a microscope. photocopies that are enlargements or reductions of the original.

2. Practice multiplying a fraction by a whole number:

   a. $\frac{1}{2} \cdot 6 = \underline{3}$

   b. $\frac{1}{3} \cdot 12 = \underline{4}$

   c. $\frac{3}{5} \cdot 10 = \underline{6}$

   d. $\frac{1}{4} \cdot 0 = \underline{0}$

Dilations, Part 1

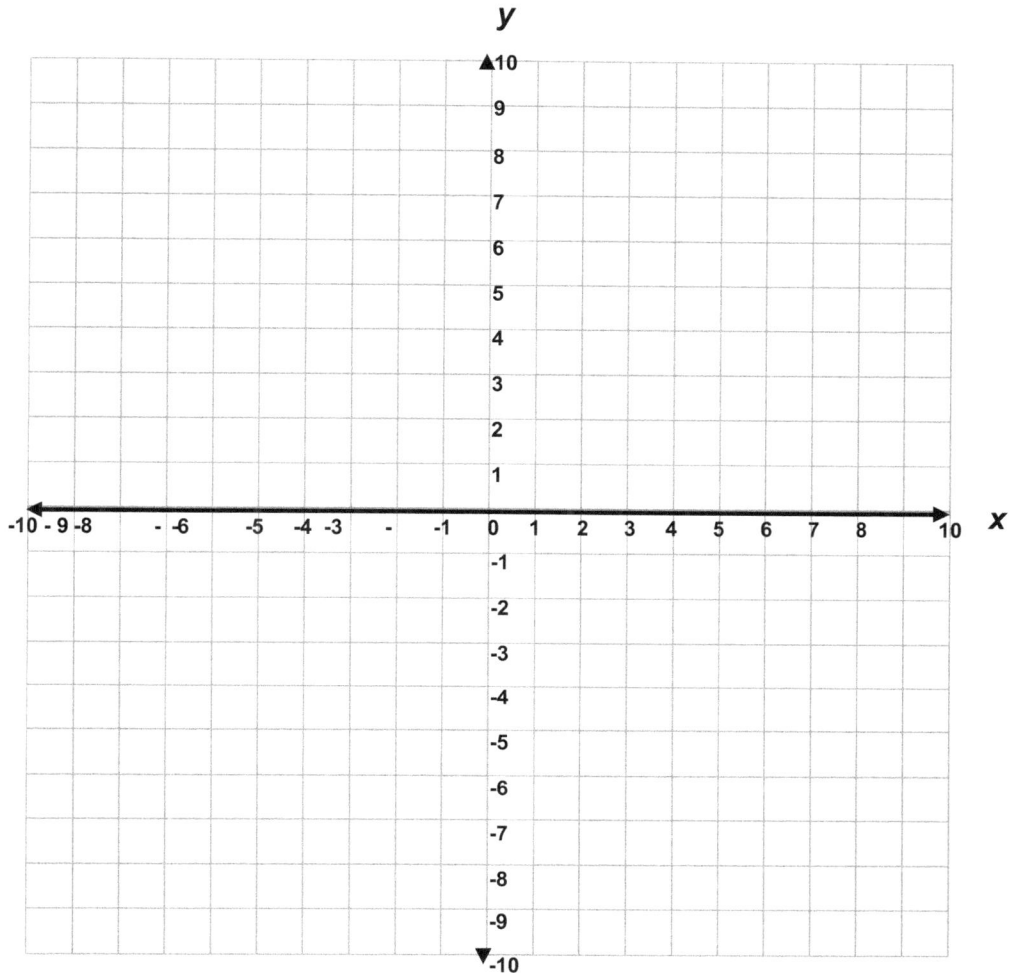

SCALE FACTOR	COORDINATES				Length of AB	Length of AD
original	A (−1, −1)	B (2, −1)	C (2, 3)	D (−1, 3)		
2	A' ( )	B' ( )	C' ( )	D' ( )		
3	A" ( )	B" ( )	C" ( )	D" ( )		

1. What happens to the size of the figure after dilating it, using a scale factor of 2?

   _____

2. What happens to the size of the figure after dilating it, using a scale factor of 3?

   _____

3. Does a dilation cause the shape to change? _____

Name: _____

Dilations, Part 2

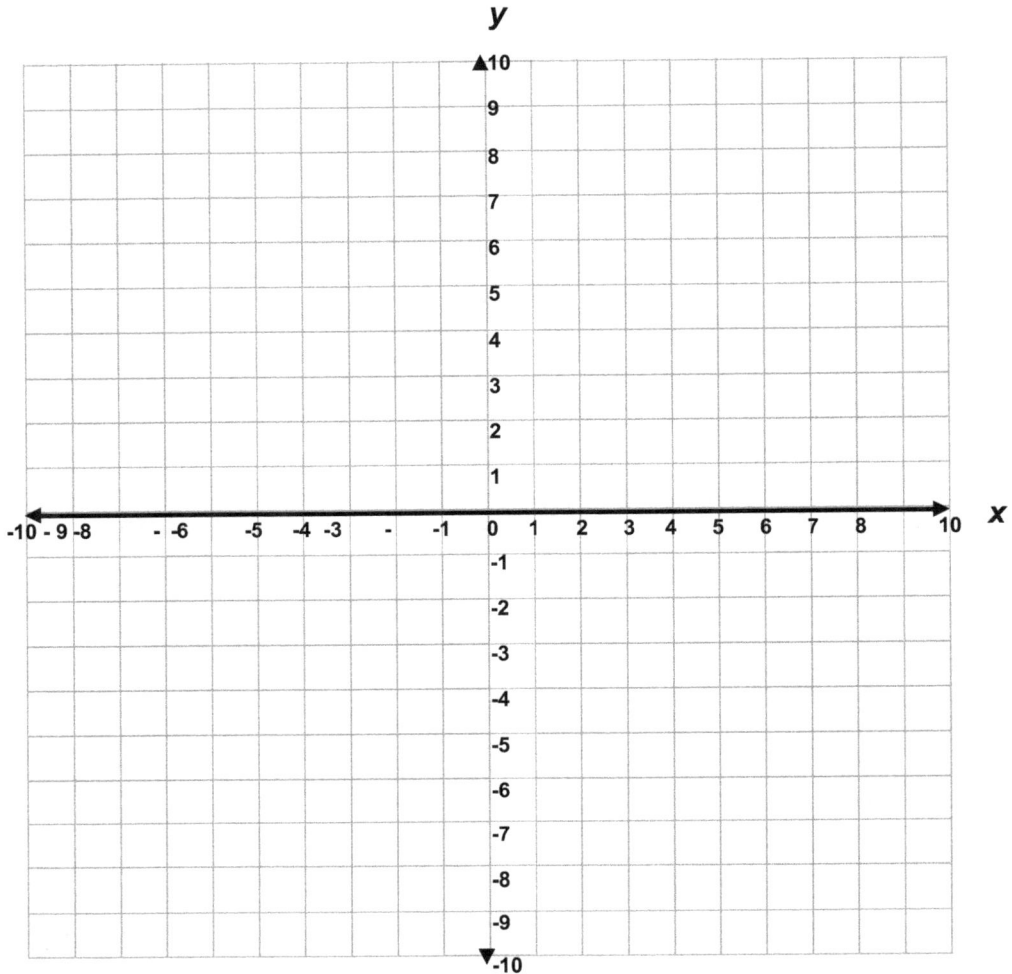

SCALE FACTOR	COORDINATES				
original	A (0, 4)	B (4, 0)	C (8, 4)	D (10, 10)	E (4, 8)
½	A' ( )	B' ( )	C' ( )	D' ( )	E' ( )

1.   What happens to the size of the figure after dilating it, using a scale factor of ½?

_____

2.   Did the shape change? _____

Dilations, Part 1

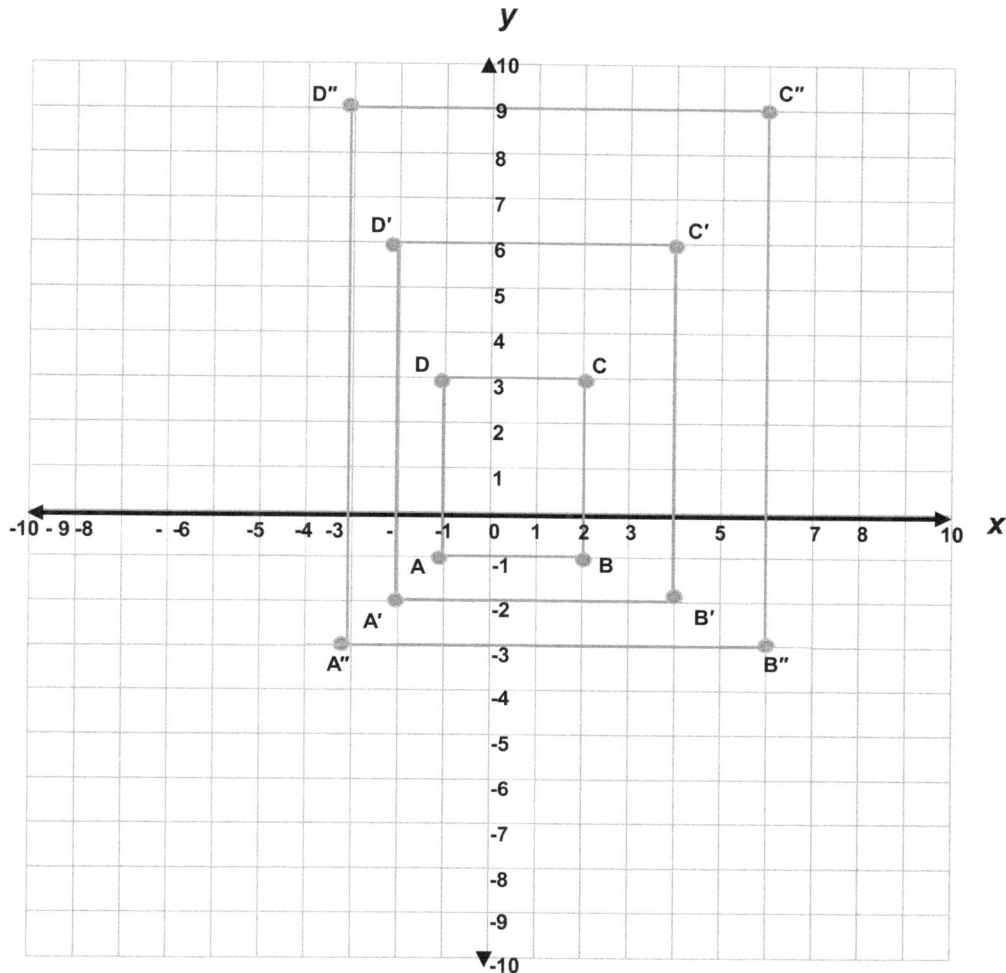

SCALE FACTOR	COORDINATES				Length of AB	Length of AD
(original)	A (−1, −1)	B (2, −1)	C (2, 3)	D (−1, 3)	3 units	4 units
2	A′ (−2, −2)	B′ (4, −2)	C′ (4, 6)	D′ (−2, 6)	6 units	8 units
3	A″ (−3, −3)	B″ (6, 3)	C″ (6, 9)	D″ (−3, 9)	9 units	12 units

1. What happens to the size of the figure after dilating it, using a scale factor of 2?

    It gets twice as large.

2. What happens to the size of the figure after dilating it, using a scale factor of 3?

    It gets three times as large.

3. Does a dilation cause the shape to change? No, only the size changes.

Dilations, Part 2

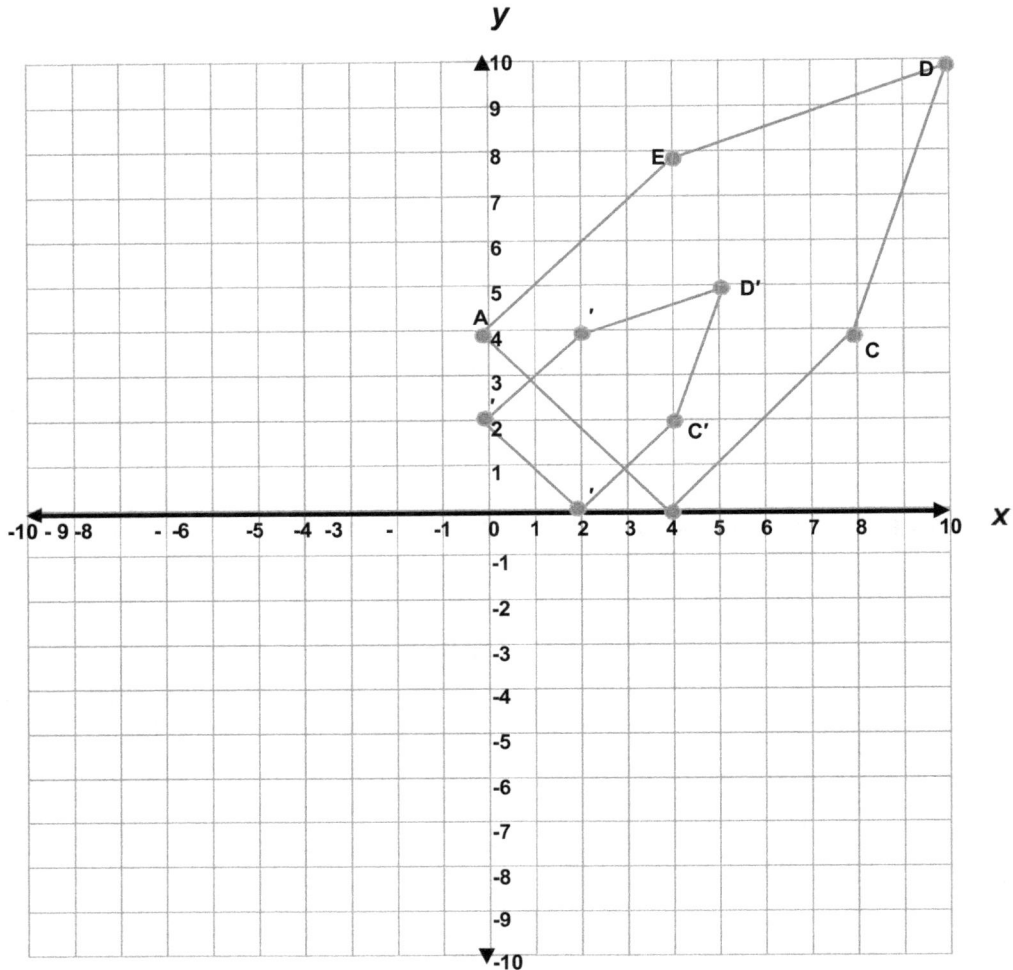

SCALE FACTOR	COORDINATES				
(original)	A (0, 4)	B (4, 0)	C (8, 4)	D (10, 10)	E (4, 8)
½	A' (0, 2)	B' (2, 0)	C' (4, 2)	D' (5, 5)	E' (2, 4)

1.   What happens to the size of the figure after dilating it, using a scale factor of ½?

_____It gets half as large._____

2.   Did the shape change? No, only the size changes._____

Reflecting on Dilations

1. How would you describe a dilation?

    _____

2. If you dilate a figure using a scale factor greater than 1, what happens to the figure?

    _____

3. If you dilate a figure using a scale factor less than 1, what happens to the figure?

    _____

4. If you were to dilate a figure using a scale factor of 1, what do you think would happen to the figure?

    _____

Name: <ins>ANSWER KEY</ins>

Reflecting on Dilations

1. How would you describe a dilation?

    <ins>It causes a figure to change in size, but its shape stays the same.</ins>

2. If you dilate a figure using a scale factor greater than 1, what happens to the figure?

    <ins>It gets bigger.</ins>

3. If you dilate a figure using a scale factor less than 1, what happens to the figure?

    <ins>It gets smaller.</ins>

4. If you were to dilate a figure using a scale factor of 1, what do you think would happen to the figure?

    <ins>It would stay the same size.</ins>

Name: _____

Reflections, Part 1

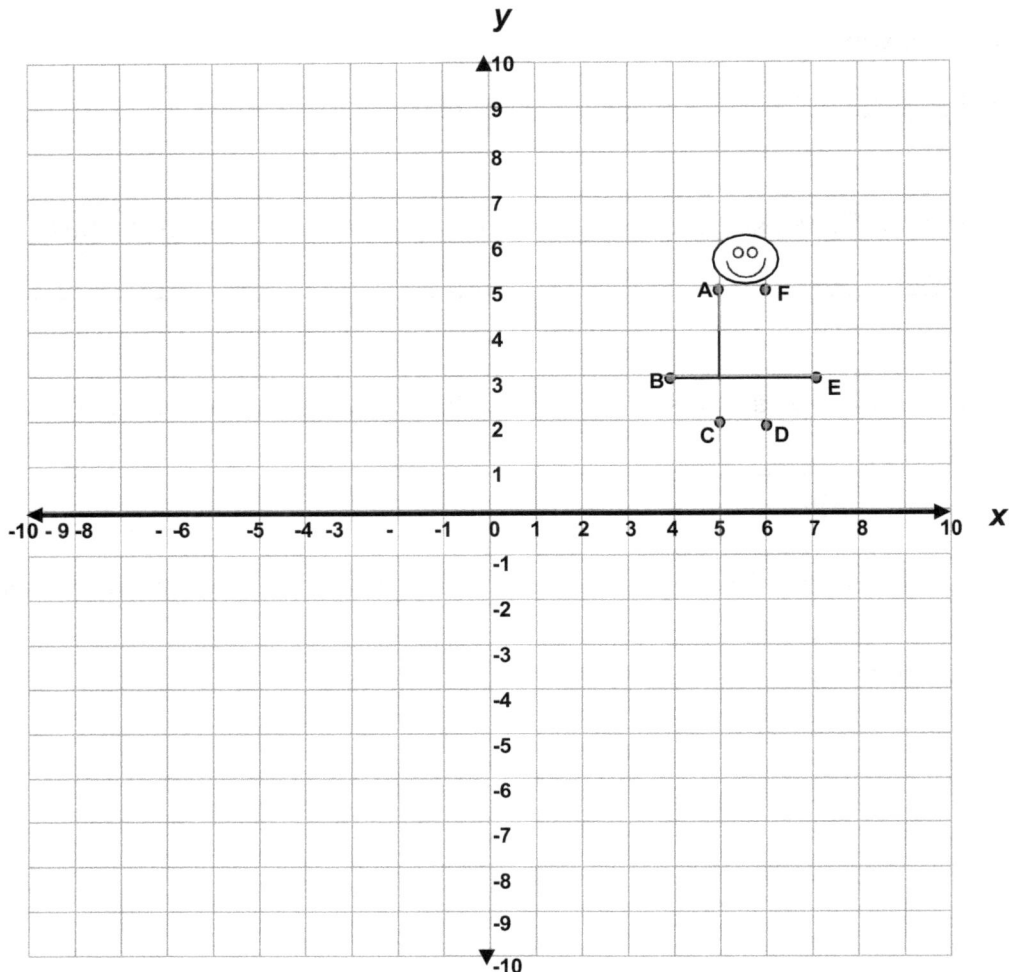

1. In the table below, write the coordinates of the figure above and its image after reflecting it across the *x*-axis.

A (	)	B (	)	C (	)	D (	)	E (	)	F (	)
A′ (	)	B′ (	)	C′ (	)	D′ (	)	E′ (	)	F′ (	)

2. Compare the sets of coordinates. What do you notice about these pairs?

_____

3. Count how many blocks each point is from the *x*-axis, and list them below:

A ___5___   B_____   C_____   D_____   E_____   F_____

A′_____   B′_____   C′_____   D′_____   E′_____   F′_____

4. The distance from the mirror line to any point on the figure is _____
the distance from the mirror line to its *reflected image*.   (greater than, less than, or equal to)

144

Scratch Paper

Name: _____

Reflections, Part 2

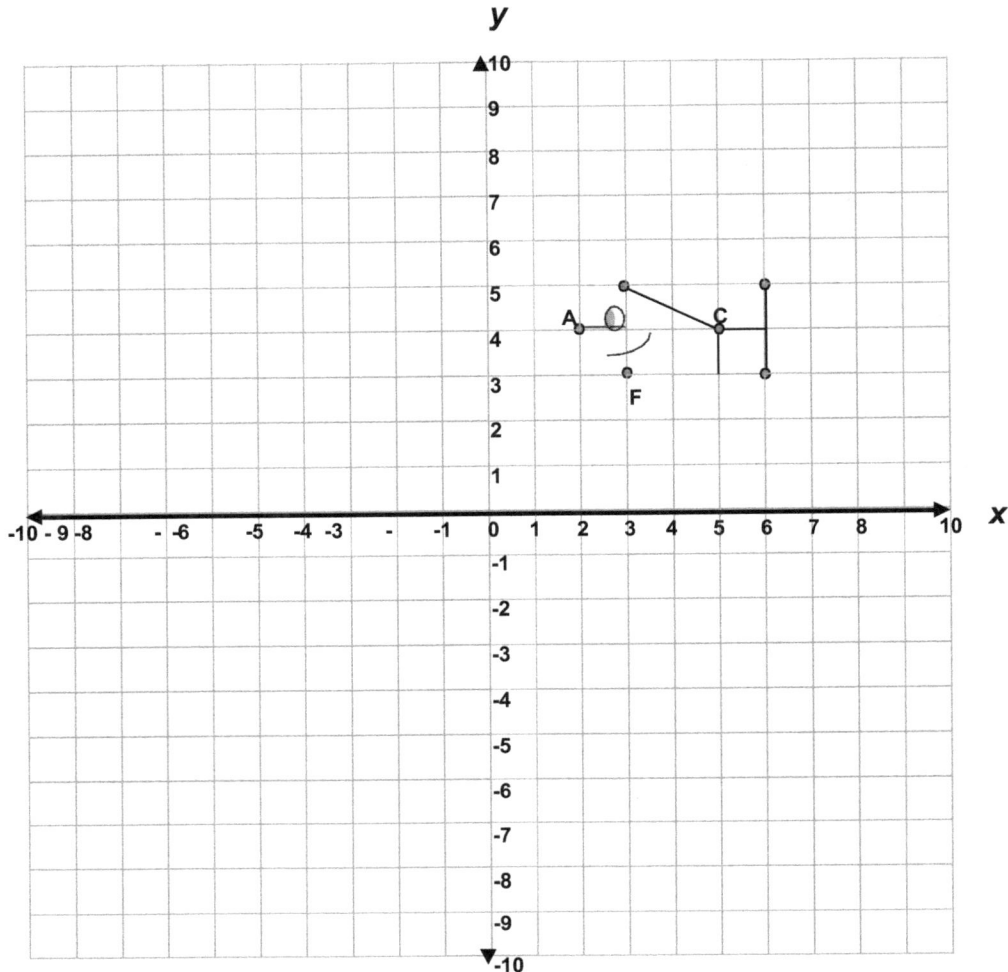

1. In the table below, write the coordinates of the figure above and its image after reflecting it across the *y*-axis.

A ( )	B ( )	C ( )	D ( )	E ( )	F ( )
A' ( )	B' ( )	C' ( )	D' ( )	E' ( )	F' ( )

2. Compare the sets of coordinates. What do you notice about these pairs?

_____

3. Count how many blocks each point is from the *y*-axis, and list them below:

A __2__   B_____   C_____   D_____   E_____   F_____

A'_____   B'_____   C'_____   D'_____   E'_____   F'_____

4. The distance from the mirror line to any point on the figure is _____
   the distance from the mirror line to its *reflected image*.   (greater than, less than, or equal to)

146

Scratch Paper

Name: ANSWER KEY

Reflections, Part 1

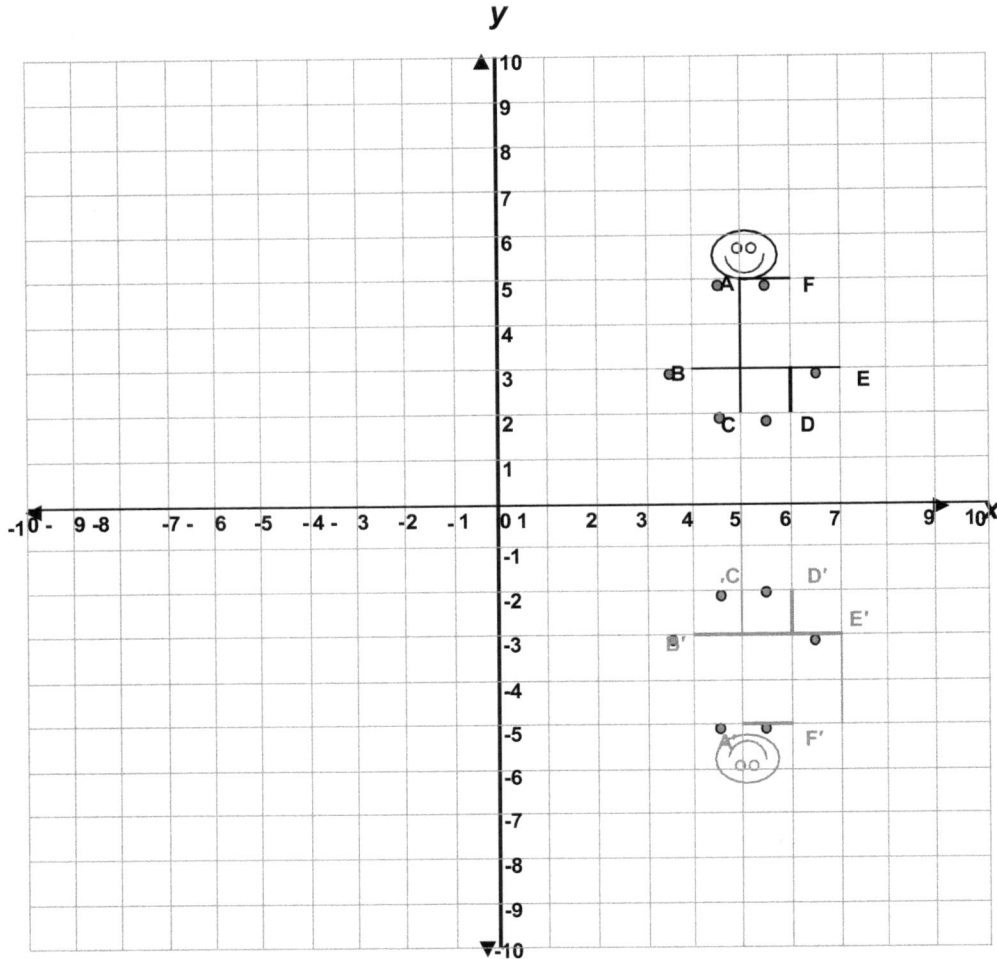

1. In the table below, write the coordinates of the figure above and its image after reflecting it across the *x*-axis.

A (5, 5)	B (4, 3)	C (5, 2)	D (6, 2)	E (7, 3)	F (6, 5)
A′ (5, −5)	B′ (4, −3)	C′ (5, −2)	D′ (6, −2)	E′ (7, −3)	F′ (6, −5)

2. Compare the sets of coordinates. What do you notice about these pairs?

    The *x*-coordinates stay the same, but the *y*-coordinates change signs.

3. Count how many blocks each point is from the *x*-axis, and list them below:

    A   5   B   3   C   2   D   2   E   3   F   5

    A′   5   B′   3   C′   2   D′   2   E′   3   F′   5

4. The distance from the mirror line to any point on the figure is ____equal to____ the distance from the mirror line to its *reflected image*.
    (greater than, less than, or equal to)

148

Name: ANSWER KEY

Reflections, Part 2

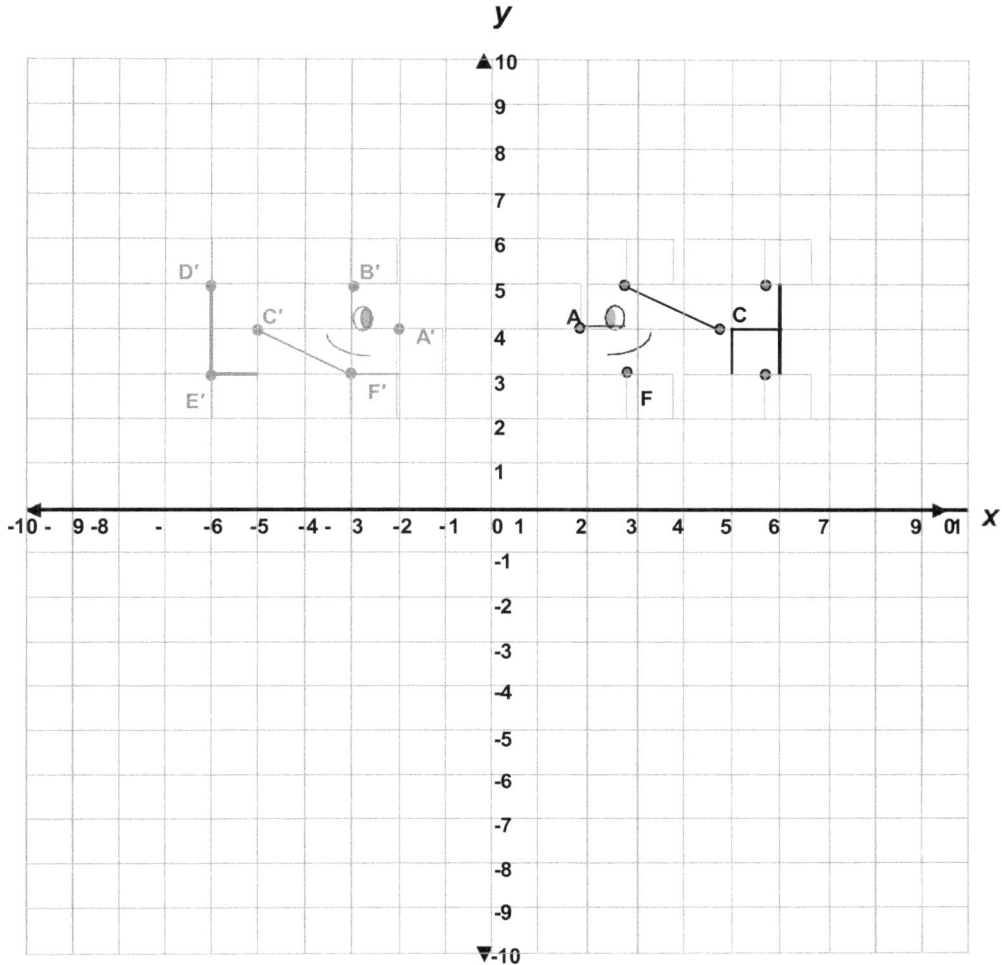

1. In the table below, write the coordinates of the figure above and its image after reflecting it across the *y*-axis.

A (2, 4)	B (3, 5)	C (5, 4)	D (6, 5)	E (6, 3)	F (3, 3)
A′ (−2, 4)	B′ (−3, 5)	C′ (−5, 4)	D′ (−6, 5)	E′ (−6, 3)	F′ (−3, 3)

2. Compare the sets of coordinates. What do you notice about these pairs?

   The *x*-coordinates change signs, but the *y*-coordinates stay the same.

3. Count how many blocks each point is from the *y*-axis, and list them below:

   A __2__   B __3__   C __5__   D __6__   E __6__   F __3__

   A′ __2__   B′ __3__   C′ __5__   D′ __6__   E′ __6__   F′ __3__

4. The distance from the mirror line to any point on the figure is ___equal to___ the distance from the mirror line to its *reflected image*.
   (greater than, less than, or equal to)

149

Reflecting on Reflections

1. How do the coordinates change when an object is reflected across the *x-axis*?

   _____

2. How do the coordinates change when an object is reflected across the *y-axis*?

   _____

Use your conclusions above to find the new coordinates for each point:

3. (4, 3) reflected across the x-axis _____

4. (−2, 1) reflected across the y-axis _____

5. (0, 0) reflected across the y-axis _____

6. (−5, −6) reflected across the x-axis _____

7. (0, 9) reflected across the y-axis _____

8. (0, 9) reflected across the x-axis _____

9. (7, 0) reflected across the y-axis _____

10. (7, 0) reflected across the x-axis _____

11. Reflect the figures below across the *x-axis*.

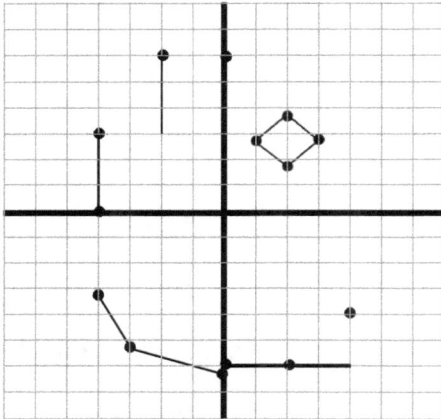

12. Reflect the figures below across the *y-axis*.

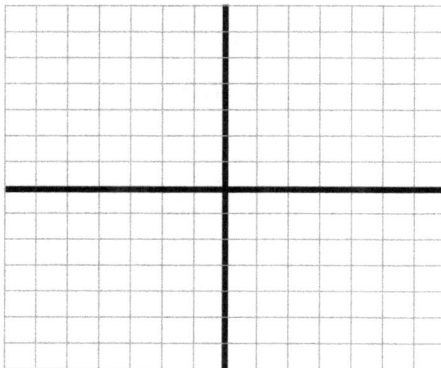

Scratch Paper

Reflecting on Reflections

1. How do the coordinates change when an object is reflected across the *x-axis*?

    The *x*-coordinate stays the same, but the *y*-coordinate changes sign.

2. How do the coordinates change when an object is reflected across the *y-axis*?

    The *x*-coordinate changes sign, but the *y*-coordinate stays the same.

Use your conclusions above to find the new coordinates for each point:

3. (4, 3) reflected across the *x*-axis (4, −3)

4. (−2, 1) reflected across the *y*-axis (2, 1)

5. (0, 0) reflected across the *y*-axis (0, 0)

6. (−5, −6) reflected across the *x*-axis (−5, 6)

7. (0, 9) reflected across the *y*-axis (0, 9)

8. (0, 9) reflected across the *x*-axis (0, −9)

9. (7, 0) reflected across the *y*-axis (−7, 0)

10. (7, 0) reflected across the *x*-axis (7, 0)

11. Reflect the figures below across the *x*-axis.

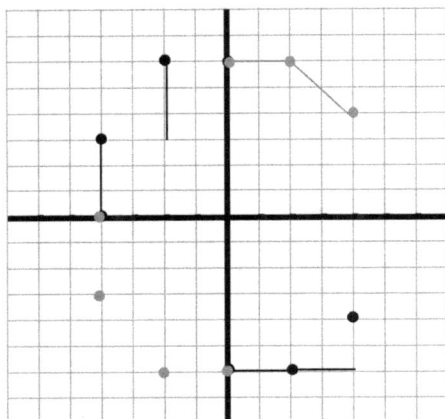

12. Reflect the figures below across the *y*-axis.

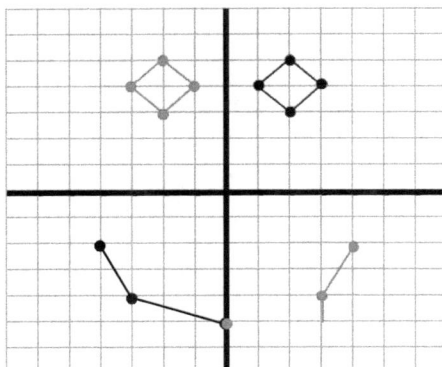

S.S. Publishing